VOLKSWAGEN TRANSPORTER/BUS

1949-67

The VW Samba Bus, displayed at the International Automobile Exhibition, April 19-29, 1951 in Frankfurt, West Germany. Two months later this chic vehicle went into production.

VOLKWAGEN TRANSPORTER/BUS

1949-67

A Documentation by Walter Zeichner

1469 Morstein Road, West Chester, Pennsylvania 19380

This volume of the Schiffer Automotive Series is devoted to the VW Transporter or VW Bus, particularly the models before the 1967 model year, recognizable by the divided front windshield. We have tried to use contemporary pictorial material to portray these vehicles as they appeared in their time. How fascinating, in particular, are those illustrations in old brochures in which the artist was hired to exaggerate a mundane transporter into a classy streamlined freighter . . . in the process of which there was sometimes so much exaggeration that it strikes us funny today. But this does not in any way detract from the artistic achievements of the graphic artists.

This book, which also shows the contemporary rivals of the VW Bus (and it certainly had some!), is not meant to be a reference work for all the technical data of this vehicle or an absolutely complete coverage of all the models built then, nor to offer repair advice or other technical information. It is not our purpose to claim to be complete—there are plenty of other books, including VW typologies, to do that. The Schiffer Automotive Series are rather picture books of automotive nostalgia and are meant to take the observer back to an era that is yet not too far back, which we remember well, with which many of us are linked by a portion of our life history. Whoever owns a pre-1967 Bus will also be able to authenticate the originality of his vehicle or find inspiration for authentic restoration...

We must not forget to express our thanks to those colleagues who were of help, especially Johannes Fleischer, Artur Westrup of the Neckarsulm auto press, Bernd Schultz of the Model Box in Frankfurt am Main, Georg Amtmann and Hans J. Klersy. The Volkswagen AG in Wolfsburg kindly gave us permission to publish historically interesting materials.

Halwart Schrader
Editor

Translated from the German by Dr. Edward Force.

Copyright © 1989 by Schiffer Publishing Ltd.
Library of Congress Catalog Number: 89-084168.

Printed in the United States of America.
ISBN: 0-88740-196-1
Published by Schiffer Publishing Ltd.
1469 Morstein Road, West Chester, Pennsylvania 19380

Please write for a free catalog.
This book may be purchased from the publisher.
Please include $2.00 postage.
Try your bookstore first.

Originally published under the title "VW Transporter/Bus" 1949-67, Schrader Motor Chronik, copyright Schrader Automobil-Bücher, Handels-GmbH, München, West Germany, ©1986, ISBN: 3-922617-09-3.

Contents

Advertising photo from VW from 1959, taken in the Austrian Alps. The winter practicality of the Bus became proverbial; in this respect, the bus equaled the VW Beetle.

Volkswagen Bulli—Almost a Legend
The Developmental History of a Universal, Useful Vehicle

When the first VW Transporters rolled off the production line in February of 1950, this was something like a sensation for all those business people who needed a medium-sized delivery . Up to this point there were-with very few exceptions-only three wheeled vehicles or station wagons developed from passenger cars. The "three pointers" were often enough models designed in the Thirties and were usually underpowered for covering long distances. They could scarcely be used at all off the road.

The new VW Transporter filled a gap, beyond all doubt. It was planned as a three-quarter-ton vehicle, had the nature of the passenger car and represented modern conceptions of motor vehicle construction. And it was reasonably priced, costing only a little more than a VW Beetle. The first of them, available exclusively as box vans, cost only 5850 Marks at the factory.

In the first informative brochures one could read that the vehicle was built in "aircraft construction"— which was then a favorite synonym for light construction in general. The VW Transporter had, in fact, no chassis in the oldtime sense, but a self-supporting body. This was a very modern design for a commercial vehicle.

Praise was won by the good handling (especially in the mountains and in winter) provided by rear-wheel drive, and by the robust and powerful four-cylinder boxer motor with air cooling, just as in the Beetle. The vehicle offered an almost equal weight distribution no matter what load it carried, as well as a carefully tuned suspension. All in all, it was an almost perfect solution to numerous transport problems.

But the "Bulli", as the Transporter soon came to be known to Continental Europeans, also had a few disadvantages. One was the poor loading conditions through the rear hatch, where the powerplant created a barrier. The sales strategists thus emphasized the wide wing doors (later a sliding door) on the sidewalk side, which was not always practical though. Then too, the position of the cab at the very front made some people think that the Bulli was very dangerous for driver and passenger in case of a collision. Finally, it was said that all the bumps in the road would be transmitted to the spinal column of the driver, who sat directly over the front axle. Obviously, the advantages outweighed the disadvantages; they included the proverbial VW reliability, the good workmanship of the vehicle and the fine service network. So the Bulli quickly rose to become the most popular vehicle in its class—in which it soon had all kinds of competition.

VW open truck in use on a building site.

Only in 1960 was the VW Bus/Transporter made with directional lights instead of levers; the 6-volt electric system was retained until 1966.

As early as March of 1950, only a month after production had begun, the so-called Kombi wagon version was introduced. With this vehicle either goods or people could be transported; two bench seats could easily be installed or removed. In addition, it was possible to add an opening fabric roof panel. In June of 1951 there then appeared the "Samba Bus", a model intended exclusively for transporting people, with an opening roof, two-tone paint, lots of chrome and luxurious interior decor. This bus could hold up to nine people. And no one could complain of poor visibility—the vehicle had, believe it or not, 21 windows! Two rows ran along the edges of the roof, a type of styling that was soon adopted by other manufacturers . . .

In August of 1952 the VW pickup truck came on the market, featuring a closed cab and a 4.2 square meter rear bed with opening side panels; it was a very popular variant that many had awaited eagerly. If weather-sensitive goods were to be transported, the truck could be equipped with a canvas rear cover and hoops.

In 1954—the year in which the hundred-thousandth Transporter left the factory on October 9—important improvements were made to the Bulli. The cylinder bore was enlarged by two millimeters, giving a displacement of 1192 cc and increasing power by 25 to 30 HP, so that the box van could now attain nearly 100 kph. The heater and seats were also improved, and a combined ignition and starter lock made its use easier.

Modifications and improvements took place every year: in 1955 all models received a full-width dashboard (formerly only in the Samba Bus) instead of a small instrument panel in front of the steering wheel, and the fuel filler was moved to the right side, over the rear axle. In addition, hydraulic shock absorbers and smaller wheels were introduced. In 1959—by which time the Transporter had been built for three years at a new factory in Hannover—the pickup truck could also be ordered with a double cab or a larger rear bed; in 1960 the old-fashioned directional signals were replaced by modern lights. The motor's power was increased to 34 HP, and automatic starting was introduced.

In 1962 the driver's seat was detached from the two (now folding) passenger seats and made adjustable; the spare wheel well was deepened. And in 1963 a 42-HP 1.5-liter motor was available optionally, good for more than 100 kph. This year's numerous innovations also included the sliding side door as an option costing 250 Marks. The rear hatch was widened, the front directional lights enlarged. And a one-ton version of the VW Bulli appeared.

A motor governor was introduced by VW in 1964, and at the same time the windshield wipers were made stronger and combined with a windshield washing system. A year later the 1.5-liter powerplant was improved to produce 44 HP, the rear hatch was given a larger window, and the front suspension gained stabilizers. As of October 1964 only the 44-HP version was available. Only in 1966 was the electric system changed from 6 to 12 volts. Now one key opened all locks, and seat belts could be added as an option.

7

Two-tone

Zweifarben-Lackierung

Naturally there were countless special versions of the VW Bulli—for a thousand and one uses. The camping buses were especially popular; there were complete, built-in sets of camping facilities—for example, by Westfalia—and a lot of extra equipment to turn the commercial vehicle into a full-value spare-time van. With such vehicles, vacationers and adventurous travelers invaded all continents. The Bulli proved to be an always reliable partner, even under the most extreme conditions, on the worst surfaces and under the most demanding use . . .

In June of 1967 the last VW Type 2 (as the Bulli was called at the factory) with a divided windshield was built, that vehicle, already nostalgic today. that did its job around the world, and to some extent still does so today, as well over 1.8 million of them were built. The oldest of them are already sought out and carefully restored by VW fans.

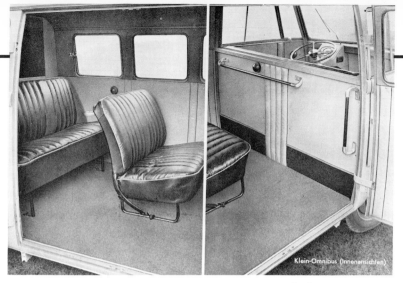

Klein-Omnibus (Innenansichten)

Small Bus (interior views)

Klein-Omnibus

Small Bus

These pictures appeared in the first brochure for the Volkswagen "Type 2", as it was called at the factory. The prototype, photographed in 1949, still had vertical louvers, thick car-type seat upholstery and a spare wheel housed vertically next to the motor. The rear wall went all the way to the roof; behind the third row of seats was a closed bulkhead.

Next two pages: Bernd Reutters was the house photographer of the Volkswagen factory. From him come these well-done, slightly idealized portrayals of the cars from Wolfsburg. The box and Kombi versions were joined by the open truck in 1952.

Motor compartment

Motorenraum

Die VW-Transporter

Die ideale Gewichtsverteilung: Fahrer vorn, Motor hinten, die Last in der Mitte im bestgefederten Raum.

Die großen Außenflächen wie geschaffen für wirkungsvolle Werbung.

The ideal weight distribution: driver in front, motor in back, the load in the middle, in the best-sprung space.

Now 4.8 cubic meters of storage space. and as before, room for 3 people in the cab.

The large exterior surfaces, as if created for effective advertising.

Jetzt 4,8 cbm Laderaum und nach wie vor Platz für 3 Personen in der Fahrerkabine.

10

Dropping side, and tailboards make loading the handy VW open truck easy.

Klappbare Bordwände, rollbare Planenteile machen das Laden leicht beim wendigen VW-Pritschenwagen.

The VW Open Truck
More than 6 square meters of total loading surface, at ramp level, completely flat platform, 4.2 square meters. Storage space between the axles, 1.9 square meters, locking, water- and dust-tight.

Der VW-Pritschenwagen

Mehr als 6 qm Gesamtladefläche. Rampenhohe, durchgehend ebene Plattform mit 4,2 qm. Tresor zwischen den Achsen 1,9 qm, abschließbar, wasser- und staubsicher.

Cover with bows for a small extra charge.

Plane mit Spriegeln gegen geringen Aufpreis.

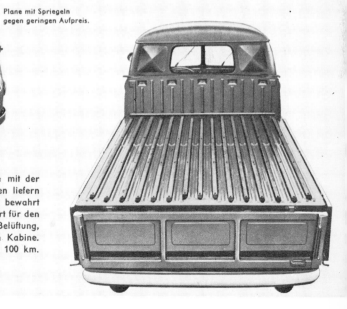

The right "speedster" for businesses that often load with handcarts at ramp level, and that must carry varying freight for many orders. The locking storage space in the best-sprung area protects especially valuable goods from strange hands. Comfort for the driver: standard heating and controllable special ventilation, as well as hinged and lowering windows in the three-seat cab. Carrying capacity 800 kg. Average fuel consumption 9.5 liters per 100 km. Highest sustained speed 80 kph with full load.

Der richtige »Flitzer« für Betriebe, die oft in Rampenhöhe mit der Karre laden, die unterschiedliches Gut in vielen Kommissionen liefern müssen. Der abschließbare Tresor im bestgefederten Raum bewahrt besonders zu schützendes Gut vor fremdem Zugriff. Komfort für den Fahrer: serienmäßige Heizung und regulierbare Spezial-Belüftung, außerdem Schiebe- und Ausstellfenster in der dreiplätzigen Kabine. Tragfähigkeit 800 kg. Durchschnittsverbrauch 9,5 Liter auf 100 km. Dauer-Höchstgeschwindigkeit 80 km/h bei voller Belastung.

11

Der VW-Kombi

dessen drei
grundsätzlich unterschiedliche
Verwendungs-Möglichkeiten ihn zu einem
tatsächlich universellen Transporter machen

The VW Kombi
Whose three basically different potential uses make it a truly universal transporter.

Viele Branchen setzen den VW-Kombi in folgenden Variationen nutzbringend ein: 1 – als Lieferwagen mit großem, hellem Laderaum und allen Vorzügen des VW-Kastenwagens. 2 – als Achtsitzer, denn im Handumdrehen sind die zwei bequemen Polsterbänke eingesetzt. Der von außen zugängliche, vergrößerte Heckraum bleibt großem Gepäck vorbehalten. 3 – als »kombinierter« Kombi – zur gleichzeitigen Beförderung von Personen und Gütern im Innenraum. Die Ausstattung ist bewußt schlicht und ganz dem wandelbaren Zweck angepaßt, denn der Laderaum soll schnell und leicht gereinigt werden können.

Many VW Kombis are put to good use in the following ways: 1—as a delivery truck with large, bright loading space and all the advantages of the VW box van. 2. as an eight-seater, for with a twist of the wrist the two comfortable upholstered seats are installed. The enlarged rear space, accessible from outside, is reserved for large luggage. 3—as "combined" Kombi—for simultaneous transport of people AND goods in the interior. The decor is deliberately simple and completely suited to changing tastes, for the loading space must be quickly and easily cleanable.

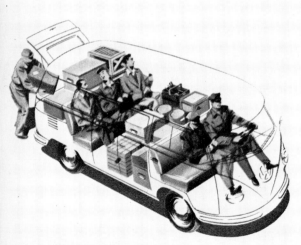

Der VW-Kombi auf Montagefahrt:
6 Personen finden bequem Platz,
dennoch bleibt der größte Teil des Innenraums
frei für Ware, Material oder Gerät.

Leicht sind
die Flügelschrauben
gelöst und die Sitze
entfernt.

The VW Kombi on a trip: 6 people have comfortable seats, yet the greatest part of interior is free for merchandise, materials or equipment.
The wing nuts are easily loosened and the seats removed.

VW-Achtsitzer

VW Eight-seater

The Kombi wagon appeared in February of 1950. One could use it optionally for transporting freight or, by installing two seat benches, turn the Bulli to an eight-seater. Above is the title page of the 1954 Kombi brochure, at right the technical data of the 30 HP Bulli.

Technisches

Motor	Luftgekühlter 4-Zylinder-4-Takt-Boxermotor im Heck des Fahrzenges. Zylinderbohrung 77 mm — Hub 64 mm — Hubraum 1192 cm³
Höchstleistung	30 PS bei 3400 U/min
Getriebe	Viergänggetriebe; 2., 3. und 4. Gang sperrreynchronisiert und geräuscharm
Fahrwerk	Einzelradaufhängung und Torsionsstabfederung vorn und hinten. Superballonreifen 6.40—15
Kraftstoffverbrauch	Normverbrauch 8,5 l/100 km · Durchschnittsverbrauch 9,5 l/100 km
Dauer- und Höchstgeschw.	80 km/h bei 3300 U/min
Steigfähigkeit	1. Gang 24%, 2. Gang 12%, 3. Gang 7,5%, 4. Gang 4%
Gewichte	Eigengewicht 1085 kg, Leergewicht 1110 kg, Nutzlast 740 kg, zulässiges Gesamtgewicht 1850 kg
Abmessungen über alles	Länge 4190 mm, Breite 1725 mm, Höhe 1940 mm
Gepäckraum-Abmessungen	Länge 700 mm, Breite 1450 mm, Höhe 800 mm — 0,8 m²
Rückwand-Ladetür	Lichte Maße: Breite 900 mm, Höhe 730 mm

Änderungen vorbehalten · Printed in Germany · A. Bagel, Düsseldorf

Technical Data

Motor: Air-cooled 4-cylinder 4-stroke boxer motor in the rear of the vehicle. Cylinder bore 77 mm—stroke 64 mm—displacement 1192 cc.

Maximum power: 30 HP at 3400 rpm

Transmission: Four-speed transmission. 2nd, 3rd and 4th gears synchronized and quiet.

Suspension: Independent suspension and torsion bars front and rear, 640—15 super-balloon tires.

Fuel consumption: Normal consumption 8.5 liters per 100 km—average consumption 9.5 liters per 100 km.

Sustained and top speed: 80 kph at 3300 rpm

Climbing ability: 24% in 1st gear, 12% in 2nd gear, 7.5% in 3rd gear, 4% in 4th gear.

Weights: Net weight 1085 kg, dry weight 1110 kg, load limit 740 kg, allowable gross weight 1850 kg.

Overall dimensions: Length 4190 mm, width 1725 mm, height 1940 mm.

Load space dimensions: Length 700 mm, width 1450 mm, height 800 mm '0.8 square meter.

Rear loading door: Dimensions: Width 900 mm, height 730 mm.

Left: the enchantingly lovely advertising graphics of the early Fifties! Thus the Bulli was portrayed as a streamlined express van . . .above and right: advertising photos from the 1951 brochure that Wolfsburg issued for the so-called Samba Bus. Whoever did not find the many windows sufficient could buy this model with an extra—large-size—fabric opening roof.

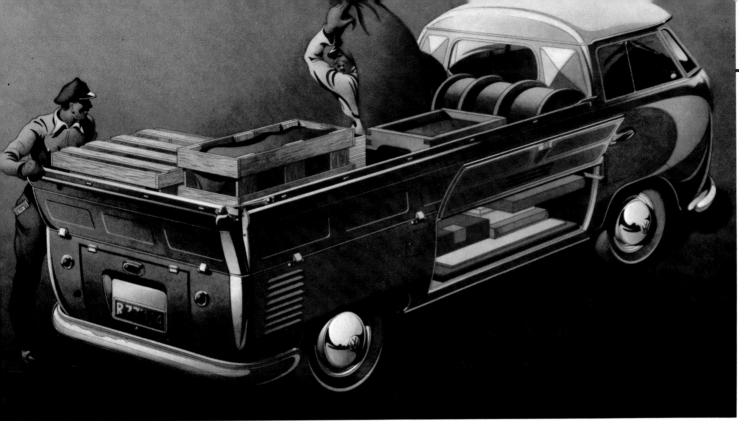

Space utilization: under the rear bed was an additional storage space of 1.55 square meter surface.

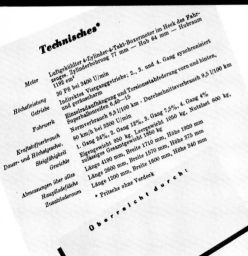

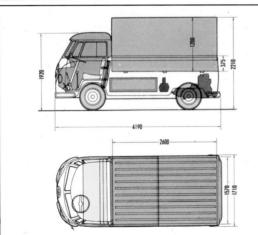

Technical Data

Motor: Air-cooled 4-cylinder 4-stroke boxer motor in the rear of the vehicle. Cylinder bore 77 mm—stroke 64 mm—displacement 1192 cc.

Maximum power: 30 HP at 3400 rpm.

Transmission: Indirect 4-speed transmission; 2nd, 3rd and 4th gears synchronized and quiet.

Suspension: Independent suspension with torsion bars front and rear, 640—15 super-balloon tires.

Fuel consumption: Normal consumption 8.5 liters per 100 km, average consumption 9.5 liters per 100 km.

Sustained and top speed: 80 kph at 3300 rpm.

Climbing ability: 24% in 2st, 12% in 2nd, 7.5% in 3rd, 4% in 4th gear.

Weights: Net weight 950 kg, dry weight 1050 kg, load limit 800 kg, allowable gross weight 1850 kg.

Overall dimensions: Length 4190 mm, width 1710 mm, height 1920 mm.

Main load area: Length 2600 mm, width 1570 mm, height 375 mm.
Additional load area: Length 1200 mm, width 1600 mm, height 375 mm.

Der VW-Kombi hat die anerkannten Vorzüge des VW-Kastenwagens,
nämlich einen nach Größe und Beladefähigkeit uneingeschränkten Nutzraum für alle Sorten von Waren
und Material, außerdem aber noch den Zugang des Tageslichtes durch die breiten Seitenfenster.
Der VW-Kombi verfügt zudem aber auch über alle Eigenschaften des VW-Achtsitzers, denn im Nu
sind die breiten Polsterbänke eingesetzt, die den Wagen in einen flotten Personentransporter verwandeln.
Und schließlich kann man — durch eine nur teilweise Verwendung der Sitze — Personen und Waren
oder Gerät in nahezu beliebiger Variation gleichzeitig unterbringen, womit die Universalität
dieses für die unterschiedlichsten Zwecke verwendbaren Wagens wirklich unüberbietbar ist.
Die Innenausstattung ist ganz und gar auf praktische Nützlichkeit ausgerichtet;
auch Ware, die Schmutz hinterläßt, braucht man nicht zu scheuen,
denn der Innenraum ist so leicht und bequem zu reinigen,
wie sein schneller Umbau zur Personenbeförderung dies erfordert.

The VW Kombi has the well-known advantages of the VW box van, namely an unlimited useful space, in terms of size and loading capability, for all sorts of goods and materials, and in addition the benefit of daylight through the wide side windows. The VW Kombi also offers all the advantages of the VW Eight-seater, for in no time the wide upholstered benches are installed, turning the van into a speedy transporter of people. And finally—by using only part of the seating—one can accommodate people and goods or equipment simultaneously in practically any desired variation, in which the universality of this vehicle, useful for the greatest variety of purposes, is really unbeatable. The interior furnishings are completely designed for parctical utility; goods that leave dirt behind need not be avoided, for the interior is so easy and handy to clean as its quick change into a carrier of people makes possible.

For thirteen years the Transporter/Bus was available only with two opening side doors, until in 1963 an optional sliding door was offered. Of course there was a version with the doors on the other side for countries with left-side traffic.

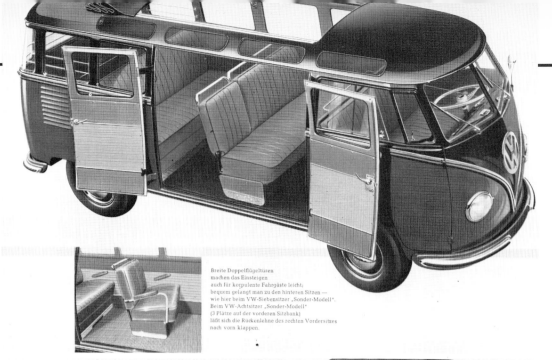

Breite Doppelflügeltüren
machen das Einsteigen
auch für korpulente Fahrgäste leicht;
bequem gelangt man zu den hinteren Sitzen —
wie hier beim VW-Siebensitzer „Sonder-Modell".
Beim VW-Achtsitzer „Sonder-Modell"
(3 Plätze auf der vorderen Sitzbank)
läßt sich die Rückenlehne des rechten Vordersitzes
nach vorn klappen.

Officially the Samba
Bus was called
"Special Model".
The windows in the
rear corners and in
the roof were made of
plexiglas. Right:
brochure pictures
(1955) of the normal
eight-seater with its
no less comfortable
furnishings.

VW-Sieben- oder Achtsitzer „Sonder-Modell"

Wide double doors make getting in easy even for corpulent passenge
one easily gets to the rear seats—as here in the VW seven-sea
"Special Model" (3 seats on the front bench), where the back of
right front seat can be folded forward.
VW seven- or eight-seater "Special Model".

Mit ihm fährt der Erfolg

Clearly—the successful businessman chose a VW! From 1954 on, the four-cylinder boxer motor produced 30 HP instead of the original 25.

Ship a lot at once and quickly.
The VW open truck has the advantage of a thoroughly flat bed at ramp height (and under it a large locking compartment in the best-sprung space between the axles). The VW box van offers two easy ways of loading: through the double side doors (from the safety of the sidewalk) or through the ramp-level rear hatch.

Schnell und viel auf einmal zuverlässig befördern

Der VW-Pritschenwagen hat den Vorteil der durchgehend ebenen Ladefläche in Rampenhöhe (darunter ein Tresorfach im best gefederten Raum zwischen den Achsen). Beim VW-Kastenwagen ist für zwei bequeme Lademöglichkeiten gesorgt: durch die seitliche Doppelflügeltür (vom gefahrlosen Bürgersteig her) und die rampenhoch angelegte

20

Der VW-Kombi vereint die Vorzüge des VW-Kastenwagens mit denen eines Personen-Transporters. Rasch sind zwei Polsterbänke eingebaut, und der Sieben- oder Achtsitzer ist fertig. Außerdem kann der Kombi bei Verwendung von nur einer Bank als geräumiges Gemischt-Fahrzeug dienen.

Three transporters in one
The VW Kombi unites the advantages of the VW box van with those of a transporter of people. Two upholstered benches are easily fitted, and the seven or eight-seater is finished. In addition, the Kombi can be used as a roomy mixed transporter by using only one bench.

The (almost) only disadvantage of the VW Bulli was that one could lock goods away not from the rear, but only from the side. But the high sales figures showed that this was not regarded as a very great handicap . . .

Underway with the whole family, carefree and independent ...
naturally in a VW seven- or eight-seater.

Spare time and business car in one—an ideal all-round automobile.
The brochure from which these pages come was published in 1955.
Now the Bulli has 15-inch wheels and a full-width dashboard.

4-cylinder 4-stroke carbureted motor in the
the vehicle.
ers: Two opposed banks of two cylinders
linder bore 77 mm, stroke 64 mm, displace-
192 cc Compression ratio: 6.6 : 1
Dropped
30 HP at 3400 rpm
speed: 7 m/s at top speed ' 80 kph (3300

ation: Pressure lubrication (geared pump)
l cooler
stem: Mechanical fuel pump
etor: Downdraft with accelerator pump
er: Oil-bath filter
g: Air-cooled by fan, automatically
ed by thermostat

single-plate dry clutch

nission
our-speed gearbox; 2nd, 3rd and 4th gears
onized and quiet
tios: 1st gear 1:3.60; 2nd gear 1:1.88; 3rd
.22; 4th gear 1:0.82; reverse gear 1:4.63

xle Drive
transmission: by spiral-geared bevel wheel,
ing gears, swing axles and spur gears to the
eels
tio: 1 : 6.2

suspension: 2 laminated torsion bars
aspension: 1 torsion bar on each side
absorbers: Front and rear double-action
pic shock absorbers
g: ZF-Ross steering, hydraulic steering

g circle: approximately 12 meters
akes: Hydraulic four-wheel brakes (Ate)
brake: Mechanical, working on the rear

4.5 K x 15, deep-bed rims
5.40-15
ase: 2400 mm
Front 1370 mm, rear 1360 mm

nance
nsumption: By DIN standards 9.5 liters per
(Open truck without rear cover: 10 liters
km).
ed and top speed: 80 kph
ng ability: 1st gear 24%, 2nd gear 12%, 3rd
%, 4th gear 4%

Motor
Bauart 4-Zylinder-4-Takt-Vergasermotor
im Heck des Fahrzeuges
Zylinderanordnung je 2 Zylinder gegenüberliegend
Maße Zylinderbohrung 77 mm - Hub
64 mm - Hubraum 1192 cm²
Verdichtungsverhältnis 6,6
Ventile hängend
Höchstleistung 30 PS bei 3400 U/min
Kolbengeschwindigkeit 7 m/s bei Höchstgeschwindigkeit
= 80 km/h (3300 U/min)
Schmierung Druckumlaufschmierung
(Zahnradpumpe) mit Ölkühler
Kraftstoff-Förderung mechanische Kraftstoffpumpe
Vergaser Fallstromvergaser
mit Beschleunigungspumpe
Luftfilter Ölbadfilter
Kühlung Luftkühlung durch Gebläse,
automatisch durch Thermostat
geregelt

Kupplung
Bauart Einscheiben-Trockenkupplung

Wechselgetriebe
Bauart Vierganggetriebe; 2., 3. und
4. Gang sperrsynchronisiert
und geräuscharm
Übersetzungsverhältnis 1. Gang 1 : 3,60 2. Gang 1 : 1,88
3. Gang 1 : 1,22 4. Gang 1 : 0,82
Rückwärtsgang 1 : 4,63

Hinterachsantrieb
Kraftübertragung durch spiralverzahntes Kegelrad-
getriebe, Kegelradausgleichsge-
triebe, Pendelachsen und Stirnrad-
untersetzung auf die Hinterräder
Übersetzungsverhältnis 1 : 6,2

Fahrgestell
Federung vorn 2 lamellierte Profil-Drehfeder-
stäbe
Federung hinten 1 Drehfederstab auf jeder Seite
Stoßdämpfer vorn und hinten doppeltwirkende
Teleskopstoßdämpfer
Lenkung ZF-Roß-Lenkung, hydraulischer
Lenkungsdämpfer
Wendekreis etwa 12 m
Fußbremse hydraulische Vierradbremse (Ate)
Handbremse mechanisch, auf die Hinterräder
wirkend
Räder 4½ K x 15, Tiefbettfelge
Reifen 6,40—15
Radstand 2400 mm
Spurweite vorn 1370 mm, hinten 1360 mm

Fahrleistungen
Kraftstoffverbrauch nach DIN 70030 9,5 l / 100 km
(Pritschenwagen ohne Plane
10 l / 100 km)
Dauer- u. Höchstgeschw. 80 km/h
Steigfähigkeit 1. Gang 24 %, 2. Gang 12 %,
3. Gang 7,5 %, 4. Gang 4 %

Technical data worth knowing

		Open without cover	Box van	Kombi	Seven- and eight-seater	Seven- and eight-seater "Special Model"
Weights:	Net weight (kg)	950	920	940	1085	1085
	Dry weight (kg)	1050	1020	1040	1110	1110
	Load limit (kg)	800	830	810	740	740
	Allowable gross weight (kg)	1850	1850	1850	1850	1850
	Number of seats	3	3	3	7/8	7/8
Overall dimensions:	Length (mm)	4190	4190	4190	4190	4220
	Width (mm)	1710	1725	1725	1725	1750
	Height (mm)	1920	1940	1940	1940	1940
Other dimensions: Double side doors:	Width (mm)	–	1170	1170	1170	1170
	Height (mm)	–	1200	1200	1200	1200
Rear door:	Width (mm)	–	900	900	900	900
	Height (mm)	–	730	730	730	730
Cargo bed height from ground (empty):	Front (mm)	980	500	500	500	500
Ground clearance:	(mm)	240	240	240	240	240
Interior dimensions: Cargo or passenger compartment: ger compartment:	Median length (mm)	2600	2700	2700	Passenger compartment for 5 to 6 persons, depending on arrangement of seats	
	Median width (mm)	1570	1500	1500		
	Median height (mm)	375	1350	1350		
	Cargo surface (m²)	4.2	–	–		
	Cargo space (m³)	–	4.8	4.8		

Und ganz zum Schluß:

Vielleicht hat sich auch bei Ihnen die Auffassung verstärkt, daß man mit dem VW-Transporter einen wirklich
tüchtigen Mitarbeiter gewinnt, der wesentlich mehr einbringt als er kostet. In diesem Zusammenhang
weisen wir noch auf die günstigen Finanzierungsbedingungen der Volkswagen-Finanzierungsgesellschaft
hin, die eine VW-Anschaffung so sehr erleichtern (bei 12 Monatsraten nur 5,5 Prozent Finanzierungsgebühr).
Auch das dichte VW-Kundendienst- und Werkstättennetz mit seinen bekannt niedrigen, allgemein verbind-
lichen Richtpreisen garantiert erhebliche finanzielle Vorteile. Wir würden uns aufrichtig freuen, Sie dem-
nächst bei uns zu einer Probefahrt begrüßen zu können und danken Ihnen für die aufmerksame Lektüre.

AUTOHAUS
Johann JANSSEN G.m.b.H.
Leer-Ostfriesl.

So erfüllt der VW-Transporter als echter Volkswagen alle Aufgaben, die ihm Industrie und Landwirtschaft, Handel und Handwerk, Behörden und Verkehrsgewerbe stellen können. Seine Vielseitigkeit hat ihn darüber hinaus zum begehrten Camping- und Familienreise-Wagen gemacht; ebenso ist er als zuverlässiges Expeditionsfahrzeug in allen Kontinenten bekannt. Kein Wunder, daß er sich sehr rasch den Weltmarkt erobert hat und im deutschen Bundesgebiet zum meistgefahrenen Transporter seiner Größenklasse geworden ist. Die verschiedenen Serien-Modelle bieten jedem Geschäftszweig den maßgerechten Wagen, zumal es auch noch über hundert erprobte Variationen mit Spezial-Inneneinrichtungen und Sonder-Aufbauten gibt, die den unterschiedlichsten Branchen und Waren angepaßt wurden.

Thus the VW Transporter, as a genuine Volkswagen, fulfills all the tasks that can be asked of it by industry and agriculture, trade and service, official and transport agencies. Its versatility has also made it a desirable camping and family-trip vehicle; likewise it has become known on all continents as a reliable expedition vehicle. No wonder that it has conquered the world market very quickly and become the most frequently used transporter of its class in the German Federal Republic. The various production models offer every type of business the right vehicle, as there are over 100 tested variations with special interior furnishings and special bodies, made for the most varied uses and loads.

Along with the basic models there were vast numbers of special versions. One could get a Bulli for practically any purpose. The factory was always receptive to special requirements.

Von allen Seiten besehen:
der VW-Pritschenwagen —
ob mit oder ohne Plane, aber
immer mit zusätzlichem „Tresor" —
ist ein gewinnendes Fahrzeug,
das sich schnell bezahlt macht.

Seen from all sides: the VW pickup truck—whether with or without a
cover, but always with its additional "safe"—is a winning vehicle that
pays for itself very quickly.

The artist here lets the storage area slide out from under the rear bed
like a drawer, providing considerable extra storage space thanks to
the high level of the bed. Notice the position of the fuel filler cap
(flap) compared to the closed models.

Dieser jüngste Spezialist unter den VW-Pritschenwagen hat sich binnen kurzer Zeit durchgesetzt. Die Kombination: offene Ladefläche — zweite Kabine, diese scheinbar so einfache Lösung eines dringend gewordenen Transportproblems, ist für eine große Reihe verschiedenartigster Betriebe und Branchen unentbehrlich geworden.

VW Pickup Truck with double cab
This newest specialist among the VW open trucks has established itself within a short time. The combination of open rear bed and second cab, this seemingly so simple solution of a transport problem that has become urgent, has become indispensible for a large number of various types of business.

In 1959 the open truck appeared with a four-to-six-seater double cab. A solution that retained its validity to our own times.

The full-width dashboard was introduced in 1955; previously only the "Special Model" was equipped with it. The modified bumpers indicate an export model. The directional lights replaced the conventional levers as of 1960.

As of 1960 the VW boxer motor produced 34 HP and had automatic starting. The Bulli still had a small rear window until 1963. One could get the rear cover and bows for the open truck only for an extra charge, but one could also mount a box body on the rear. So the Bulli really could be used universally. The base price of the pickup truck in 1960 was only 5725 Marks; cover and frames cost an extra 250 Marks.

Individually dropping walls make the rear bed easy to reach from three sides.

Covers and frames (for a slight extra charge) can be mounted or demounted with a twist of the wrist.

Einzeln klappbare Bordwände machen die Ladefläche von drei Seiten bequem zugänglich.

Plane und Spriegel (gegen geringen Aufpreis) sind im Handumdrehen auf- oder abmontiert.

Für Sonderaufbauten ist der VW-Pritschenwagen das geeignete Fahrzeug. Hier nur ein Beispiel von vielen: VW-Pritschenwagen mit Kofferaufbau.

The VW pickup truck is the right vehicle for special bodies. Here is just one example of many: VW pickup truck with cabinet rear body.

1. Sun visor
2. Storage shelf
3. Defroster ducts
4. Brake cylinder
5. Telescopic shock absorber
6. Torsion bars
7. Ventilation
8. Loading area light
9. Gearbox
10. Heater duct
11. Rear axle
12. Spur gear
13. Fuel tank
14. Fuel filler
15. Oil-bath air filter
16. Distributor
17. Fuel pump
18. Carburetor
19. Generator
20. Battery
21. Rear hatch

1 Sonnenblende

2 Ablage-Etage

3 Entfrosterdüsen

4 Radbremszylinder

5 Teleskopstoßdämpfer

6 Drehfederstäbe

7 Frischbelüftung

8 Laderaumleuchte

9 Getriebe

10 Heizschlauch

11 Hinterachse

12 Stirnrädervorgelege

13 Kraftstoffbehälter

14 Kraftstoff-Einfüllstutzen

15 Ölbad-Luftfilter

16 Zündverteiler

17 Kraftstoffpumpe

18 Vergaser

19 Lichtmaschine

20 Batterie

21 Rückwand-Ladetür

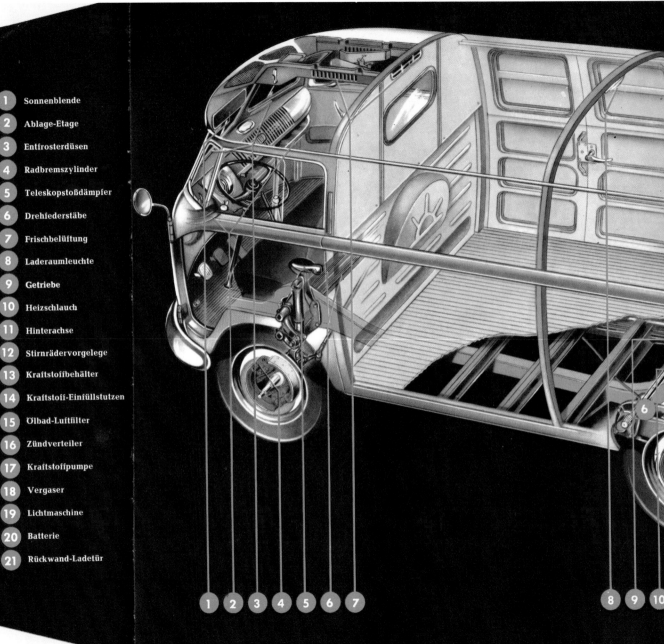

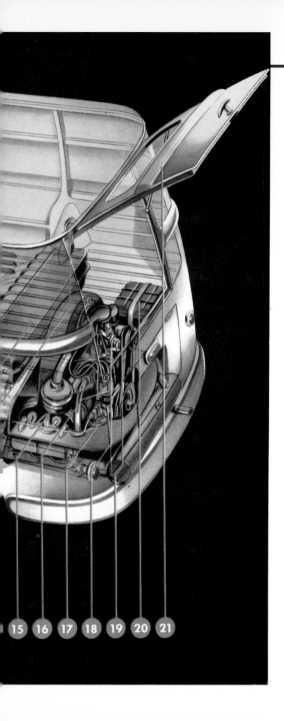

Left: An excellent portrayal of the VW box van of 1962. Above: the 1192 cc motor that was supplied in 30-HP form from 1954 to 1960, then (until 1964) with 34 HP. In 1963-64 one could optionally get the 1493 cc motor with 42 HP; this powerplant was standard as of October 1964 (1965 to 1967 with 44 HP).

Der *VW-CAMPING-WAGEN*

The VW Camper Wagon

Camping-the magic word of the early sixties! It was soon proved that the Bulli was an ideal camper, and the normal Kombi could be turned into a caravan without great expense (1960).

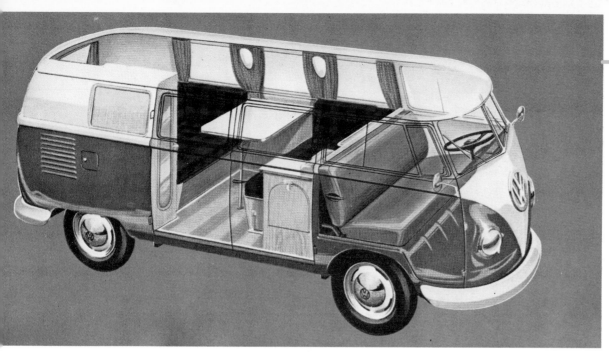

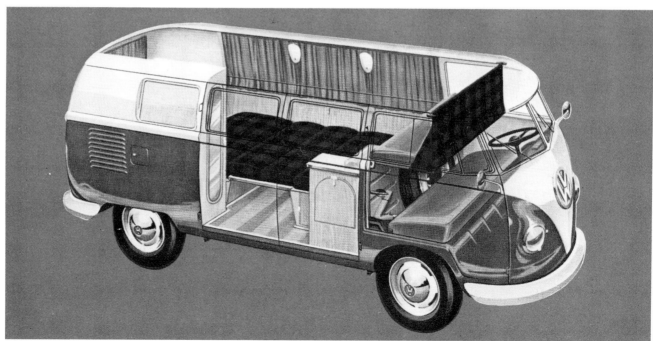

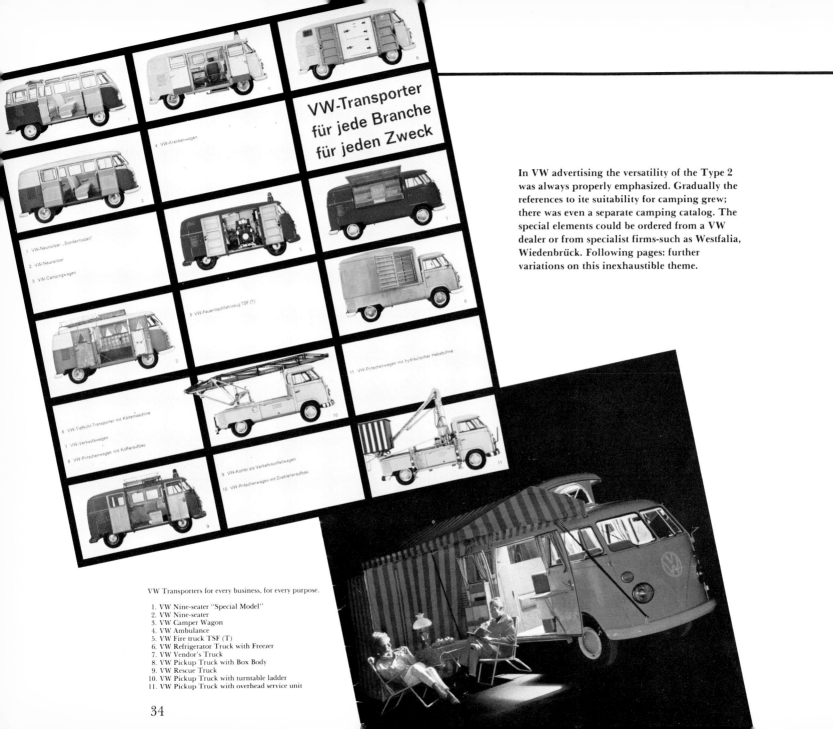

VW-Transporter
für jede Branche
für jeden Zweck

4 VW-Krankenwagen

1 VW-Neunsitzer „Sondermodell"

2 VW-Neunsitzer

3 VW-Campingwagen

5 VW-Feuerlöschfahrzeug TSF (T)

6 VW-Tiefkühl-Transporter mit Kältemaschine

7 VW-Verkaufswagen

8 VW-Pritschenwagen mit Kofferaufbau

9 VW-Kombi als Verkehrsunfallwagen

10 VW-Pritschenwagen mit Drehleiteraufbau

11 VW-Pritschenwagen mit hydraulischer Hebebühne

In VW advertising the versatility of the Type 2 was always properly emphasized. Gradually the references to ite suitability for camping grew; there was even a separate camping catalog. The special elements could be ordered from a VW dealer or from specialist firms-such as Westfalia, Wiedenbrück. Following pages: further variations on this inexhaustible theme.

VW Transporters for every business, for every purpose.

1. VW Nine-seater "Special Model"
2. VW Nine-seater
3. VW Camper Wagon
4. VW Ambulance
5. VW Fire truck TSF (T)
6. VW Refrigerator Truck with Freezer
7. VW Vendor's Truck
8. VW Pickup Truck with Box Body
9. VW Rescue Truck
10. VW Pickup Truck with turntable ladder
11. VW Pickup Truck with overhead service unit

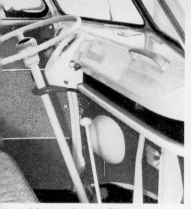

...ine und Armaturenbrett mit Ablage

Klapptisch an vorderer Flügeltür und an Isolierbox

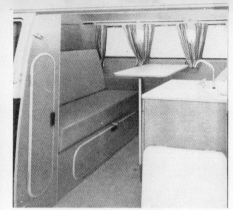

Hintere Polster-Sitzbank, darunter Staukasten, zwei Wandleuchten

...k mit Klapptisch, dahinter klappbarer Eßtisch

Regal an hinterer Flügeltür

Vorn: Kleiderschrank, an Seitenwand hinten: Eckschrank

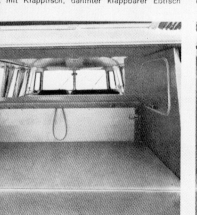

...mmi-Polster (Teil der Liegefläche), ...äscheschrank

Liegefläche für zwei Erwachsene

Mehrausstattung: Direkter Zugang zum Wohnraum

Driver's cab and dashboard with shelf

Folding table on front side door and on kitchen box

Rear upholstered seat bench, with storage space below it, two wall lights

Icebox with folding table; folding dining table behind it

Shelves on rear side door

Front: clothes closet, rear: corner cabinet on side wall

Foam rubber upholstery (part of the bed); right: bureau

Bed area for two adults

Additional equipment: direct access to the living quarters

Serienmäßige Ausstattung:

Diese Einrichtung wurde geschaffen, um VW-Sieben-, VW-Neunsitzer oder VW-Kombi als Campingwagen ausstatten zu können. Jedes Element des Anbau-Systems kann einzeln bestellt und mühelos — nach Herausnahme der Sitze — in den Fahrgastraum eingebaut und auch ebenso bequem entfernt werden.

Folgende Ausstattungsteile sind lieferbar:

Spiegel und Regalfächer.
Sisal-Teppich im Wohnraum.
Elektrische Deckenleuchte (30 Watt) mit Kabel, Stecker und Steckdose.
Drei Plastik-Wasserkanister (je 10 l Inhalt).
Hängematte für ein Kinderbett.
Dachgepäckträger.
Klapptisch an der Flügeltür.
Seitenwand-Verkleidung des gesamten Wohnraumes mit Sperrholz.

Standard Equipment

This arrangement was created to make it possible to equip a Seven-seater, VW Nine-seater or VW Kombi as a Camper Wa.. Every element of the additional system can be installed individ.. and easily—after removing the seats—built into the passenger and removed just as easily.

The following equipment is available:

Mirrors and drawers.

Sisal carpet in living quarters.

Electric ceiling light (3o watt) with cable, plug and socket.

Three plastic water canisters (each of 10-liter capacity).

Hammock for use as a child's bed.

Roof luggage rack.

Folding table on one side door.

Sidewall covering of the whole living quarters with plywood.

1. Upholstered seat bench
2. Upholstered seat bench
3. Folding table
4. Clothes closet with mirror
5. Two-door bureau with storage drawers
6. Icebox with utensil drawer
7. Toilet cabinet with wash basin, mirror and drawers
8. Children's beds, driver's seat and hammock
9. Beds for adults
10. Luggage space

Upholstered seats with folding table.

Clothes closet with mirror.

Bureau with drawers.

Wall cabinet combination:

l. Icebox (approx. 55-liter capacity) with two drawers and a bo.. wet or dry ice.

2. Utensil drawer.

Toilet cabinet.

Curtains with rods for all windows (removable in cab).

In Pegulan form over the motor compartment.

Polyester folding roof or Westfalia raising roof.

Large outside tent with extra room.

Awning equipment for large tent.

Small outside tent.

Gasoline stove.

Curtain with rod between cab and living quarters.

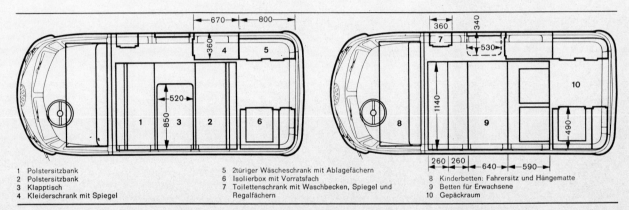

1 Polstersitzbank
2 Polstersitzbank
3 Klapptisch
4 Kleiderschrank mit Spiegel

5 2türiger Wäscheschrank mit Ablagefächern
6 Isolierbox mit Vorratsfach
7 Toilettenschrank mit Waschbecken, Spiegel und Regalfächern

8 Kinderbetten: Fahrersitz und Hängematte
9 Betten für Erwachsene
10 Gepäckraum

Polster-Sitze mit Klapptisch.
Kleiderschrank mit Spiegel.
Wäscheschrank mit Ablagefächern.
Wandschrank-Kombination:
1 - Isolierbox (ca. 5 l Inhalt) mit zwei Fächern und einer Schale für Naß- oder Trockeneis,
2 - Vorratsfach.
Toilettenschrank.
Gardinen mit Stangen für sämtliche Fenster (im Fahrerhaus abknöpfbar).

Über dem Motorraum in Pegulan-Ausführung.
Polyester-Faltdach oder Westfalia-Hubdach.
Großes Vorzelt mit Nebenraum.
Sonnendachzubehör für großes Vorzelt.
Kleines Vorzelt.
Benzinkocher.
Vorhang mit Stange zwischen Fahrerhaus und Wohnraum.

36

ine und Armaturenbrett mit Ablage

Hubdach, Dachgalerie, Klapptische an beiden Flügeltüren

Isolierbox mit Eisschale

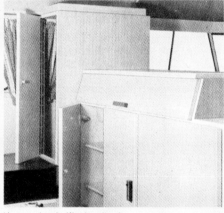

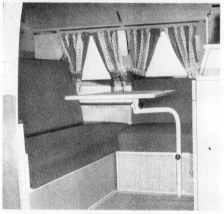

chrank, darüber Spüle, Anrichte, Kocher

Vorratsschrank, Kleiderschrank

Schwenkbarer Eßtisch und seitliche Polsterbank

olstersitzbank

Hängeschrank für Wäsche im Heck des Fahrzeuges

Liegefläche für zwei Erwachsene

Driver's cab and dashboard with storage shelf

Raising roof, roof gallery, folding tables on both side doors

Icebox with ice bowl

Dish closet, sink above it, kitchen cabinet, cabinet for cooker

Storage cupboard, clothes closet

Folding dining table and upholstered side seat

Upholstered rear seat

Hanging cabinet for clothing in the rear of the vehicle

Bed area for two adults

37

The high-roof van was introduced by VW in 1963 (upper left). Now and then one also saw vans with double doors on both sides. The maximum load varied according to body type from 580 (ambulance) to 830 kg (box van).

Fast, handy and economical as a sedan, the VW Minibus offers considerably more space. Seven to nine people find all the comforts that make even hours-long cross-country trip in any season of the year a pleasure. The external form of the vehicle, with its becoming two-tone combinations, is just as pleasing as the solid interior furnishings. One sits comfortably soft and cozy on upholstered benches with lavish imitation leather coverings. Swinging, sliding and lowering windows, together with the finely regulated special ventilation system and the quickly working full-space heating, with its heating ducts into the driver's cab, provide an interior climate always at the right temperature. In the VW seven-seater the middle seat bench has two seats; in the nine-seater with two middle benches the back of the seat nearest the door folds forward, so that one can also get to the rear seat without hindrance. A shelf on the bulkhead of the cab, armrests with foam rubber filling, handholds, ashtrays and the tasteful overall decor of the interior complete the furnishings. A sunroof is available optionally (at extra price); it opens quickly and can be bolted fast at any position. The driver's compartment is also furnished in a homey, friendly way. This is true above all of the comfortable upholstered seats, which are divided for driver and passenger in the seven-seater. The backs can be adjusted to three different inclinations. The driver's technical comfort is contributed to by: easy-to-see instrument panel with speedometer, fuei gauge and handily arranged controls, easy-to-hold steering wheel (with a steering lock that is simultaneously the ignition lock), sun visor, large storage shelf and handhold in front of the passenger seat.

As of mid-1963 the Bulli was also available with the load limit of a ton. This version was equipped with strengthened axles and brakes, as well as 700-14 tires. At the same time, the rear hatch was widened considerably, and the vehicle was given larger front directional lights. Above is a 1962 model.

Cutaway view of a Samba Bus from the end of 1962. Shortly afterward the rear corner windows were eliminated. The "Special Model" with its roof windows was produced until the autumn of 1967. Especially chic: chromed luggage rack on the rear storage surface.

1. Vent window
2. Special fresh air system
3. Sliding window
4. Sun roof
5. Roof windows
6. Hinged window
7. Luggage space
8. Chromed luggage rack
9. Handhold
10. Armrests
11. Adjustable seat back
12. Spare wheel holder with shelf
13. Padded sun visor
14. Warm air heating
15. Telescopic shock absorber
16. Fuel tank
17. Independent suspension with torsion bars

VW-Transporter

helfen gut

Geld verdienen

Das Automobil ist für die gesamte Wirtschaft zu einem unentbehrlichen Ge̲
brauchsfahrzeug geworden. Seine Anschaffung macht sich rasch bezahlt, we̲
es zuverlässig schnelle Lieferungen ermöglicht, neue Absatzbezirke erschließ̲
einen besseren Kundendienst garantiert und damit größere Umsätze und Ver̲
dienstchancen sichert. Entscheidend beim Kauf bleibt allerdings, daß man sic̲
nicht „übernimmt", sondern einen vernünftig dimensionierten Wagen wählt, de̲
den betrieblichen Notwendigkeiten auch räumlich am besten entspricht, als̲
groß genug ist und stets voll genutzt werden kann. Erst dann ist eine wirklic̲
rationelle Lösung des Transportproblems erzielt; erst dann können Fahrzeug̲
zu gewinnbringenden Mitarbeitern werden. Und das eben zeichnet die VW̲
Transporter aus! Ganz und gar auf Wirtschaftlichkeit und Zweckmäßigkeit aus̲
gerichtet, dabei auf Grund sorgfältiger Marktstudien ständig verbessert un̲

VW Transporters help to earn money
The automobile has become an indispensable, useful vehicle for the
whole economy. Its purchase quickly pays for itself, because it makes
possible reliable quick delivery, opens new sales territory, guarantees
better customer service and thus assures greater sales and chances to
earn money. Decisive in purchasing, to be sure, is that one does not
"overdo it", but chooses a vehicle of reasonable dimensions which
corresponds best to the spatial needs of the business, thus is big enough
and always can be fully utilized. Only then is a really rational solution
of the transport problem achieved; only then can vehicles become
fellow workers who bring in profits. And that is exactly what makes
the VW Transporter stand out! Designed completely for economy and
utility, and constantly improved and constructively developed on the
basis of meticulous market studies,—their construction and technical
conception have just that "golden mean" that a rational movement of
goods, equipment or people demands. That means: they provide,
along with excellent handling characteristics and low maintenance
costs, a remarkably large usable space. The perfected technology
corresponds to the solid furnishing; it also particularly meets the

onstruktiv weiterentwickelt — haben sie in Bauart und technischer Konzeption enau jenen „goldenen Schnitt", den eine rationelle Schnell-Beförderung von ütern, Geräten oder Personen verlangt. Das heißt: sie bringen neben vorzüg- hen Fahreigenschaften und niedrigen Betriebskosten einen erstaunlich großen utzraum mit. Der ausgereiften Technik entspricht die solide Ausstattung; sie ommt auch den Komfortwünschen der Fahrer und Begleiter besonders ent- egen, die täglich auf ihr Fahrzeug als engstem Berufsgefährten angewiesen sind. o erfüllt der VW-Transporter als echter Volkswagen alle Aufgaben, die ihm dustrie und Landwirtschaft, Handel und Handwerk, Behörden und Verkehrs- ewerbe stellen können; seine Vielseitigkeit hat ihn darüber hinaus zum be- hrten Camping- und Familienreise-Wagen gemacht. Kein Wunder, daß er h sehr rasch den Weltmarkt erobert hat und im deutschen Bundesgebiet zum

meistgefahrenen Transporter seiner Größenklasse geworden ist. Die verschie- denen Serien-Modelle bieten jedem Geschäftszweig den maßgerechten Wagen, zumal es auch noch über hundert erprobte Variationen mit Spezial-Innenein- richtungen und Sonder-Aufbauten gibt, die den unterschiedlichsten Branchen und Waren angepaßt sind. Sicherlich wollen Sie nun den VW-Transporter ein- mal persönlich testen — selbstverständlich ohne Kosten und Risiko und ganz unverbindlich für Sie. Sagen Sie es bitte Ihrem VW-Händler! Er schickt Ihnen gern einen „Ermächtigungsschein" zu, wie er im Muster auf der Schlußseite des Katalogs zu sehen ist. Dieser Ermächtigungsschein berechtigt Sie nach Termin- vereinbarung zu einer ausgiebigen Erprobung des VW-Transporters, und zwar mit Ihrer eigenen Ware als Ladegut und, wenn gewünscht, auch ohne Begleitung eines VW-Mannes. Unbeeinflußt können Sie sich dann selber Ihr Urteil bilden.

comfort wishes of the driver and colleagues who come to know their vehicle as their closest working companion. Thus the VW Transporter, as a true Volkswagen, fulfills all the tasks that industry and agriculture, trade and service, officialdom and transportation can ask of it; its versatility has also made it a desirable camping and family touring car. No wonder that it has conquered the world market very quickly and become the most frequently used transporter of its class in the Federal Republic of Germany. The verious production models offer every type of business the right vehicle, as there are over a hundred tested variations with special equipment and bodies, suited to the most varied businesses and loads. Surely you will now want to test the VW Transporter personally—naturally without cost, risk or obligation for you. Please tell your VW dealer! He will gladly send you a "certificate of empowerment", an example of which is shown on the last page of the catalog. This certificate allows you, after making an appointment, to make a thorough test of the VW Transporter, naturally with your own merchandise as its load and, if you wish, without being accompanied by a VW representative. Without being influenced, you can then make up your own mind.

Illustration from a Bulli catalog of 1963. Interested parties were invited to borrow a test vehicle with a "certificate of empowerment" . . .(see also page 45).

43

MOTOR

Type: 4-cylinder 4-stroke carbureted motor in the rear of the vehicle

Cylinders: 2 opposed banks of 2 cylinders each

Size: cylinder bore 77 mm—stroke 64 mm—displacement 1192 cc

Compression ratio: 7.0: 1

Valves: Dropped

Performance: 34 HP at 3600 rpm

Piston speed: 7.68 m/s at 3600 rpm

Lubrication: Pressure lubrication (geared pump) with oil cooler

Fuel system: Mechanical fuel pump

Carburetor: Solex 28 PICT downdraft carburetor with automatic choke and accelerator pump

Air filter: Oil-bath filter

Cooling: Air-cooling by fan, automatically regulated by thermostat

Clutch: Single-plate dry clutch

TRANSMISSION

Type: Four-speed gearbox, synchronized and quiet

Gear ratios: 1st gear 3.80, 2nd gear 2.06, 3rd gear 1.22, 4th gear 0.82, reverse 3.88

REAR AXLE DRIVE

Power transmission: By spiral-geared bevel wheel and equalizer, swing axles and spur gears to the rear wheels

Ratio: 5.73

SUSPENSION

Front suspension: Two laminated profiled torsion bars

Rear suspension: One torsion bar on each side

Shock absorbers: Double-acting front and rear telescopic shock absorbers

Steering: Ross steering, hydraulic damper

Turning circle: Approx. 12 meters

Foot brakes: Hydraulic four-wheel brakes

Hand brake: Mechanical, working on the rear wheels

Wheels: 4.5 K x 15, deep-bed rims

Tires: 6.40-15

Wheelbase: 2400 mm

Track: Front 1370 mm, rear 1360 mm

Fuel tank: 40 liters, including 5 liters reserve

PERFORMANCE

Fuel consumption: By DIN 70,030 9.5 liters per 100 km (open truck with cover 10.4 liters per 100 km)

Sustained and top speed: 95 kph (open truck with cover 90 kph)

Climbing ability: 1st gear 26%, 2nd gear 13.5%, 3rd gear 7%, 4th gear 4%

General: Jack and tools under the driver's seat, spare wheel behind the seat back

MOTOR

Bauart	4-Zylinder-4-Takt-Vergasermotor im Heck des Fahrzeuges
Zylinderanordnung	je 2 Zylinder gegenüberliegend
Maße	Zylinderbohrung 77 mm — Hub 64 mm — Hubraum 1192 cm³
Verdichtung	7,0
Ventile	hängend
Höchstleistung	34 PS bei 3600 U/min
Kolbengeschwindigkeit	7,68 m/s bei 3600 U/min
Schmierung	Druckumlaufschmierung (Zahnradpumpe) mit Ölkühler
Kraftstoff-Förderung	mechanische Kraftstoffpumpe
Vergaser	Solex Fallstromvergaser 28 PICT mit Startautomatik und Beschleunigungspumpe
Luftfilter	Ölbadfilter
Kühlung	Luftkühlung durch Gebläse, automatisch durch Thermostat geregelt
Kupplung	Einscheiben-Trockenkupplung

WECHSELGETRIEBE

Bauart	Vierganggetriebe; sperrsynchronisiert und geräuscharm
Übersetzungen	1. Gang 3,80 2. Gang 2,06 3. Gang 1,22 4. Gang 0,82 Rückwärtsgang 3,88

HINTERACHSANTRIEB

Kraftübertragung	durch spiralverzahntes Kegelradgetriebe, Kegelradausgleichgetriebe, Pendelachsen und Stirnraduntersetzung auf die Hinterräder
Übersetzung	5,73

FAHRGESTELL

Federung vorn	2 lamellierte Profil-Drehfederstäbe
Federung hinten	1 Drehfederstab auf jeder Seite
Stoßdämpfer	vorn und hinten doppeltwirkende Teleskop-stoßdämpfer
Lenkung	Roß-Lenkung, hydraulischer Lenkungsdämpfer
Wendekreis	etwa 12 m
Fußbremse	hydraulische Vierradbremse
Handbremse	mechanisch, auf die Hinterräder wirkend
Räder	4½ K × 15, Tiefbettfelge
Reifen	6,40—15
Radstand	2400 mm
Spurweite	vorn 1370 mm, hinten 1360 mm
Kraftstoffbehälter	40 Liter, davon 5 Liter Reserve

FAHRLEISTUNGEN

Kraftstoffverbrauch	nach DIN 70 030 9,5 l/100 km (Pritschenwagen ohne Verdeck 10 l/100 km, Pritschenwagen mit Verdeck 10,4 l/100 km)
Dauer- u. Höchstgeschw.	95 km/h (Pritschenwagen mit Verdeck 90 km/h)
Steigfähigkeit	1. Gang 26 %, 2. Gang 13,5 %, 3. Gang 7 %, 4. Gang 4 %
Allgemeines	Werkzeug und Wagenheber unter der Fahrerbank, Reserverad hinter der Banklehne

		Pickup truck w/o cover	Pickup truck with cover	Large-space pickup truck w/o cover	Large-space woodpickup truck w/o cover	Double-cab pickup truck w/o cover	Double-cap pickup truck with cover	Box van
Weights:	Dry weight (kg)	1065[1]	1100[1]	1115[1]	1135[1]	1165[1]	1190[1]	1035[1]
	Maximum load (kg)	800	765	750	730	700[3]	675[3]	830
	Allowable gross weight (kg)	1865	1865	1865	1865	1865	1865	1865
	Number of seats	3	3	3	3	6	6	3
Overall dimensions:								
	Length (mm)	4290	4290	4290	4300	4290	4290	4280
	Width (mm)	1750	1750	2020	1980	1750	1750	1750
	Height (mm)	1920	2210	1920	1920	1920	2210	1940
Other dimensions:								
Double side	Width (mm)	–	–	–	–	–	–	1170
doors width	Height (mm)	–	–	–	–	–	–	1200
Rear hatch	Width (mm)	–	–	–	–	–	–	900
	Height (mm)	–	–	–	–	–	–	730
Bed height from ground								
	At front (mm)	980	980	980	980	980	980	500
	Ground clearance (mm)	240	240	240	240	240	240	240
Interior dimensions:								
Load or pas-	Median length (mm)	2600	2600	2600	2720	1755[4]	1755[4]	2700
senger space	Median width (mm)	1570	1570	1880	1850	1570[4]	1570[4]	1500
	Median height (mm)	375	1200[6]	375	400	375[4]	1200[6]	1350
	Cargo bed (m²)	4.2	4.2	5	5	2.75[4]	2.75[4]	–
	Cargo space (m³)	–	–	–	–	–	–	4.8
Dimensions of cargo space under bed:								
	Length (mm)	1200	1200	1200	1200	–	–	–
	Width (mm)	1600	1600	1600	1600	–	–	–
	Height (mm)	340	340	340	340	–	–	–
	Cargo surface (m²)	1.9	1.9	1.9	1.9	–	–	–
	Cargo space (m³)	0.65	0.65	0.65	0.65	–	–	–

Footnotes: 1. Including driver. 2. Including driver and all seats. 3. With 5 empty seats (460 kg on bed, 435 kg with cover). 4. With 7/8 seating. 5. Areas only for rear bed surface cargo space in second cab (with seats removed). 6. Light cover height over bed.

Volkswagen-Transporter sind . . .

zügige Kletterer:	bis 26 % Steigfähigkeit mit voller Last
bescheidene Spritschlucker:	mit 40 l Tankinhalt etwa 400 km Fahrt
unermüdliche Marathonläufer:	95 km/h Dauergeschwindigkeit
behende Lücken-Nutznießer:	rasches Parken auf kleinstem Raum
wetterfeste Freiluft-Parker:	eine Garage ist überflüssig
geschäftstüchtige Mitarbeiter:	sie verdienen mehr, als sie kosten

Finanzierung leicht gemacht:

Die Anschaffung eines VW-Transporters wi... die günstigen Finanzierungsbedingungen de... wagen-Finanzierungsgesellschaft sehr erleich... Vorteil, den man bei Bedarf natürlich ge... nimmt. Bei einem auf zwölf Monatsraten la... Vertrag beträgt die Finanzierungsgebühr nur 6...

Volkswagen Transporters are . . .

Capable climbers: up to 26% climbing ability with full load

Modest fuel users: some 4000 km range with 40-liter tank capacity

Tireless marathon runners: 95 kph sustained speed

Handy small-space parkers: easy to park in smallest space

Weathertight outdoor parkers: a garage is superfluous

Businesslike colleagues: they earn more than they cost

Financing made easy:

The purchase of a VW Transporter is made much easier... favorable financing terms of the Volkswagen Fin... Organization; an advantage that one naturally notices wh... needed. For a contract running for twelve months the finance c... only 6%.

In 1964 the Bulli's motor gained a governor. The Boxer, now made in the millions, was, to be sure, not very sensitive to overrevving, but the governor also resulted in fuel saving.

Weiterkommen mit VW-Transporter

Go farther with a VW Transporter

In 1962 VW advertised with "only" 1000 service shops in the Federal Republic. Right: the "Certificate of Empowerment" for an in-service test of the Transporter, a successful sales joke.

The close-mesh VW service network includes more than 1000 contract service stations in the Federal Republic and more than 3000 VW agencies outside the country.

Das engmaschige VW-Kundendienstnetz umfaßt mehr als 1000 Vertragswerkstätten im Bundesgebiet und über 3000 VW-Vertretungen im Ausland.

We hereby place at the disposal of Mr. Mrs. Corp. a VW for a test drive and test loading for the purpose of firsthand testing for the firm.

On one of the days from to you may have the use of the VW Transporter for hours, according to your wishes. The vehicle may be picked up by you or a representative identified to us by name by presenting this certificate; at the end of the test drive we request that you return it to us at once. The option is possible of having the vehicle driven by one of our drivers during the road test.

The cost of fuel, taxes, liability and comprehensive insurance as well as any other costs are not charges to you in the case of orderly utilization while you test the vehicle.

During the test the vehicle may not be used on salaried trips, not loaned, rented or otherwise turned over to a third party.

It is expressly assured that you are under no obligation to buy on account of the test drive.

———————————————————————

(Place and date) (Signature of the VW dealer)
This page is intended for the tester.

Ermächtigungs-Schein

Hiermit stellen wir Herrn/Frau/Firma

_____

einen VW-_____
für eine Probefahrt und Probebeladung zum Zwecke der Eignungsprüfung für den Betrieb zur Verfügung.

An einem der Tage vom _____ bis _____
können Sie den VW-Transporter _____ Stunden Ihren Wünschen entsprechend einsetzen. Der Wagen kann durch Sie oder einen uns namhaft gemachten Beauftragten gegen Vorlage dieses Ermächtigungsscheines abgeholt werden; nach Beendigung der Probefahrt bitten wir, ihn uns sogleich wieder zuzustellen. Auf Wunsch besteht die Möglichkeit, den Wagen von einem unserer Fahrer bei der Eignungsprüfung fahren zu lassen.

Kosten für Betriebsstoff, Steuer, Haftpflicht- und Vollkasko-Versicherung sowie sonstige Kosten entstehen Ihnen bei ordnungsgemäßer Benutzung nicht, wenn Sie den Wagen erproben.

Das Fahrzeug darf während der Prüfung nicht für Lohnfuhren verwendet, nicht verliehen, vermietet oder sonstwie einem Dritten überlassen werden.

Es wird ausdrücklich versichert, daß Ihnen aus der Erprobung keinerlei Verpflichtungen zum Kauf erwachsen.

_____ _____
(Ort und Datum)   (Unterschrift des VW-Händlers)

Dieses Blatt ist für den Interessenten bestimmt.

VW-FEUERLÖSCHFAHRZEUG TSF (T)

VW FIRE TRUCK
TSF (T)

The VW Transporter was soon offered to communities for uses including those of fire departments. There were even separate, detailed brochures printed with fire departments in mind.

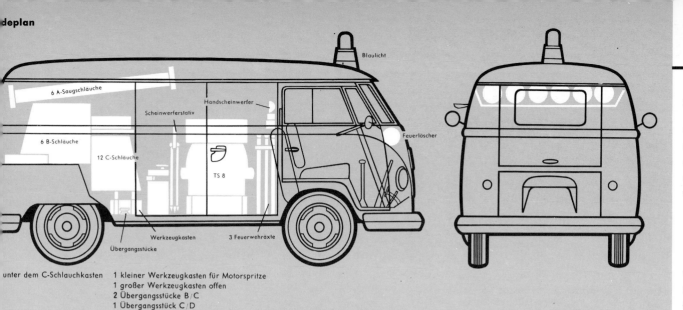

deplan

Blaulicht

6 A-Saugschläuche

Scheinwerferstativ

Handscheinwerfer

Feuerlöscher

6 B-Schlauche

12 C-Schlauche

TS 8

Werkzeugkasten

3 Feuerwehräxte

Übergangsstücke

unter dem C-Schlauchkasten 1 kleiner Werkzeugkasten für Motorspritze
1 großer Werkzeugkasten offen
2 Übergangsstücke B/C
1 Übergangsstück C/D

Weidenschutzkorb, Fangleinen

Schlauchhalt.

Kupplungsschlüssel

Kasten für Zubehör

Kübelspritze 10 l

Sani-Kasten

Standrohr 2 x B

1 B
3 C -Strahlrohre

Spaten

Schlüssel für Unterflurhydrant

Lampen

Schlüssel für Überflurhydrant

Brechstange

augkorb

Sammelstück A.—2 B

Dreiverteiler

Klappsitz

Diese Zeichnung ist nur ein Vorschlag für den Beladeplan. Einzelheiten über Konstruktion und Einbau der Innenausstattung können den „Baurichtlinien für Löschfahrzeuge" entnommen werden. Die Baurichtlinien werden herausgegeben vom Fachnormenausschuß Feuerlöschwesen (FNFW) in Verbindung mit der Arbeitsgemeinschaft Feuerschutz (AGF) und der Vereinigung zur Förderung des Deutschen Brandschutzes e.V. Sie können bezogen werden vom Fachnormenausschuß Feuerlöschwesen, Berlin W 15, Uhlandstraße 175.

Loading diagram

blue light

6A suction hose

Spotlight standard Hand spotlight Fire extinguisher

6B Hose 12C Hose

Couplings Tool chest 3 fire axes

Equipment under the C hose box

   1 small tool box for motor nozzle

   1 large tool chest, open

   2 couplings B/C

   1 coupling C/D

Wicker basket, ropes Equipment box

Hose holder 10-liter spray tank

Coupling box First aid kit Shovel

2 x B water cannon

Hydrant wrench Lights

1 B, 3 C Nozzles Crowbar Hydrant wrench

A suction coupling A-2 B Collector

Three-way divider

Folding seat

This drawing is only a suggestion for a loading arrangement. Details of construction and installation of the interior equipment can be taken from the "Construction Guidelines for Fire Vehicles". The guidelines are available from the Special Office for Firefighting (FNFW) in connection with the Firefighting Workers' Organization (AGF) and the Association for the Promotion of German Firefighting Inc. They can be obtained from the Special Office for Firefighting, Berlin W 15, Uhlandstrasse 175.

How such a firefighting vehicle by VW could be equipped is shown by the illustrations from a 1960 catalog. The Bulli was also used by the thousands as a rescue (ambulance) vehicle.

48

...n an extensive advertising campaign, the Volkswagen Works tried in 1964 to find the oldest Bulli still in existence. They discovered a vehicle from 1950 with chassis number 20-000-44. That was thus the 44th Type 2 produced. At right is the story that VW later published, along with the life stories of other veterans, as a result of this contest.

Das ist der älteste VW-Transporter.

Wir fanden ihn in Waiblingen in Württemberg. Er hat das polizeiliche Kennzeichen WN - NX 26. Er gehört dem Landmaschinenbauer Karl Widmann.

Von den vielen sehr alten VW-Transportern, die wir gefunden haben,

ist er der älteste: Fahrgestell-Nummer 20-000 44. Das heißt: Er ist der vierundvierzigste VW-Transporter, den wir gebaut haben. (Erstmals zugelassen am 18. April 1950.)

Am 9. Dezember 1959 hat ihn Karl Widmann gekauft. Aus achter Hand.

Unter dem flachen Dach der kleinen Werkstatt in Waiblingens Neustädter Straße 72 hängt ein schlichtes Schild mit der Aufschrift „Fahrzeugbau Widmann".

Karl Widmann, erfindungsreicher Schwabe, hat hier tatsächlich Fahrzeuge gebaut. Selbstentwickelte Zweiradschlepper und Landmaschinen. Viele seiner Kunden sind kleine Bauern in der Umgebung Stuttgarts. Für sie baut er heute Kleinbandsägen.

Widmann ist sein eigener Konstrukteur, Mechaniker, Schlosser und Kundendienstmann. Er liefert seine Erzeugnisse frei Haus. Dabei hat ihm der VW-Transporter gute Dienste geleistet.

Manchmal hatte er eine vier Zentner schwere Kleinbandsäge im Kasten. Dazu noch Werkzeug, Material, Kleinkram. Alles zusammen eine ausgewachsene Ladung. Und dann noch eine elf Zentner schwere Landmaschine im Schlepp.

Karl Widmann hat mit seinem VW-Transporter so ziemlich alles gefahren, was sich da hineinpacken läßt.

Er hat auch Eisenträger, Rohre und Sand transportiert. Weil er sich, so ganz nebenbei, noch ein Sägewerk und eine Fischerhütte aufgebaut hat. Da hat er auch Holz, Steine, Zement transportiert. Und junge Forellen samt Futter.

Und eine Kuh. Aus Gefälligkeit.

Oder einen Sechser-Pflug.

Oder eine Sau. Pardon. (In der Landwirtschaft geht das in Ordnung.)

VW-Großhändler Hahn in Stuttgart betrachtet das alles mit etwas gemisch-

ten Gefühlen. Denn Karl Widmann ist ein schlechter, aber schon ein ganz „schlechter" Kunde.

Erst kauft er einen gebrauchten VW-Transporter und fährt ihn viele Jahre. Und dann kommt er nicht mal zum Kundendienst. „Die paar Kleinigkeite, des wäre doch g'lacht", sagt er. „Wo i doch do mei Werkzeig hab'. Und die 'Ersatzteil' kriag mr au so günstig..."

Ja — und wenn man meint, nun

endlich ist der alte Kasten reif, nun muß er einen neuen kaufen...

... dann gewinnt er einen. Als Sieger der Aktion „Wer hat den ältesten VW-Transporter?".

„Des is aber au d's erschte Mal, daß i in mei'm Lebe was g'wonne hab."

Das ist es ja, was wir immer sagen. Fahren Sie VW-Transporter. Sie gewinnen dabei.

Karl Widmann aus Waiblingen hat inzwischen seinen fabrikneuen VW-Kastenwagen freudestrahlend übernommen. „Der isch fascht z' schad' zom Schaffe..." sagt er.

Aber jetzt, in diesem Augenblick, trägt er vielleicht schon ein Kalb über Stock und Stein. Nur so aus Gefälligkeit.

This is the oldest VW Transporter.

We found it in Waiblingen, in Württemberg. It has the license number WN—NX 26. It belongs to farm machine builder Karl Widmann.

Of the many very old VW Transporters that we have found, it is the oldest: Chassis number 20-000 44. That means: It is the forty-fourth VW Transporter that we built. (First registered on April 18, 1950.)

On December 9, 1959 Karl Widmann bought it. Eighth-hand.

Under the flat roof of the small workshop in Waiblingen, Neustädter Strasse 72, hangs a simple sign with the inscription "Fahrzeugbau Widmann".

Karl Widmann, an inventive Swabian, has really built motor vehicles here. Two-wheeled tractors and farm machines of his own design. Many of his customers are small farmers in the Stuttgart area. For them he builds small bandsaws today.

Widmann is his own constructor, mechanic, locksmith and customer service representative. He delivers his products to his customers free of charge. In the process, the VW Transporter has given him good service.

Sometimes it carries a four-hundred-pound small bandsaw in its body. As well as tools, materials, this and that. Altogether a full load. And then too, an eleven-hundred-pound farm machine in tow.

Karl Widmann has carried just about everything in his VW Transporter that can be packed into one.

He has also carried railroad ties, pipes and sand. Because he has also opened a sawmill and a fish hatchery on the side. So he has also transported wood, stones, cement. And young trout and their food.

And a cow. To do a favor.

Or a six-row plow.

Or a sow. Pardon us. (In agriculture that is quite normal.)

VW Wholesaler Hahn in Stuttgart observes all this with somewhat mixed feelings. For Karl Widmann is a bad—a really, truly "bad" customer.

First he buys a used VW Transporter and drives it for many years. And then he doesn't even come in for service. "A few little things, that would be laughable," he says. "When I have my own tools, after all. And I can get spare parts cheap..."

Yes, and when you think the old heap is finished now, now he must buy a new one...

... then he wins one. As the winner of the "Who has the oldest VW Transporter?" contest.

"This is the first time in my life I've ever won anything."

That's what we always say. Drive the VW Transporter. You'll win thereby.

Meanwhile Karl Widmann of Waiblingen has joyfully taken possession of his factory-new VW Box Van. "It is almost too nice to use..." he says.

But now, at this moment, he may be carrying a calf over sticks and stones. Just doing a favor.

This vehicle also came from the very first production run and ranked among those oldtimers that VW located in 1964. Fourteen-year-old Bullis were then something special— today there are survivors that have been running for more than 20 years!

. . . and another VW
Bus from 1950, which
was also honored in
the VW Oldtimer
brochure. Whether it
still exists today is
questionable . . .

With its variety of
different models, the
Volkswagen Works
were ahead of any
competitors. This Bulli
advertising photo
appeared late in the
summer of 1963.

Der 66er Transporter ist derselbe wie der 60er. Richtig?

(Falsch.)

54

1960

1,2-Liter-Motor mit 34 PS.

1960
1.2 liter motor with 34 HP.

In a very thorough brochure, VW showed in 1966 what developments the Transporter had gone through in the course of the past six years. The motor was just one of the many examples that made clear how much had changed . . .

1966

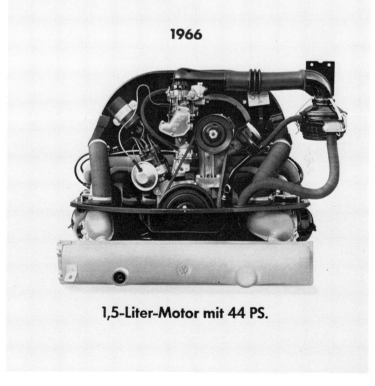

1966
1.5 liter motor with 44 HP.

1,5-Liter-Motor mit 44 PS.

1960

Kleine Hecktür.
Kleines Heckfenster.
Kleine Blinker.

1960
Small rear hatch.
Small rear window.
Small directional lights.

Two more double pages from the comparison brochure of 1966. Right: now there was also, optionally, the (later standard) sliding door. A year before, the 750-kg-carrying version had been given up. From now on the Bulli was equipped with attachments for seat belts, and the one-key system and a 12-volt battery were new.

1966
Large rear hatch.
Large rear window.
Large directional lights.

1966

Grosse Hecktür.
Grosses Heckfenster.
Grosse Blinker.

1966

Ebenfalls mit grosser Flügeltür.

1966
Likewise with big double doors.
Or with a new sliding door.

Oder mit neuer Schiebetür.

9 Detailverbesserungen, die der 60er noch nicht hatte.

Erstens trennten wir den Fahrersitz von der Sitzbank und sorgten dafür, daß Sitz und Lehne des Fahrers verstellbar sind.

Zweitens gaben wir dem Fahrerhaus zwei gepolsterte Sonnenblenden.

Drittens bauten wir eine Kraftstoffuhr ein.

Viertens brachten wir Heizluftdüsen für Fahrer und Beifahrer an.

Fünftens erhielten beide Fahrerhaustüren Türfeststeller.

Sechstens vergrösserten wir die Scheibenwischerarme, gaben ihnen zwei Wischergeschwindigkeiten und sorgten dafür, dass sie nach dem Abschalten wieder automatisch in Ruhestellung gehen.

Siebtens wurden die Türgriffe feststehend und bekamen Druckknopfbetätigung.

Achtens erhielten alle VW-Transporter serienmässig eine Scheibenwaschanlage.

Neuntens bauten wir asymmetrisches Abblendlicht ein und erhöhten so nochmals die Sicherheit.

Further detail improvements with which the 1966 model was advertised. At right are the complete technical data of that year.

9 detail improvements, that the '60 did not yet have

First we separated the driver's seat from the seat bench and made the driver's seat and back adjustable.

Second, we put two padded sun visors in the cab.

Third, we added a fuel gauge.

Fourth, we installed hot air ducts for driver and passenger.

Fifth, both cab doors gained holders.

Sixth, we enlarged the windshield wiper arms, gave them two speeds and made them come automatically to their resting position when shut off.

Seventh, the door handles were made immobile with push-button operation.

Eighth, all VW Transporters got a standard windshield washer system.

Ninth, we installed asymmetrical low-beam headlights, thus increasing safety again.

Technical Data

Motor

Four-cylinder, four-stroke boxer motor in the rear of the vehicle. Bore x stroke 83 x 69 mm. Displacement 1493 cc. Compression ratio 7.5 : 1. Produces 44 DIN HP at 4000 rpm, or 53 SAE HP at 4200 rpm. Maximum torque 10.4 mkg at 2000 rpm (DIN). Average piston speed 9.20 m/s at 4000 rpm. Pressure lubrication with oil cooler. Thermostatically regulated air-cooling by fan. Downdraft carburetor with automatic choke, accelerator pump and oil-bath air filter. Intake air and mixture pre-warming.

Electric System

Battery 12-volt/45 Ah. Generator 30 Amp. maximum with early charging. Combines ignition-starter lock with starter repeat prevention. Headlights with asymmetrical low-beam light. Directional lights with automatic shutoff. Two interior lights (one in the pickup truck). Windshield wipers with automatic return and two speeds. Flashers combined with hand dimmer switch.

Transmission

By single-plate dry clutch, fully synchronized four-speed gearbox. Equalizing gears, swing half-axles and spur gears to the rear wheels. Ratios for 1st through 4th gears 3.80/2.06/1.26/0.82, reverse 3.88, spur gearing 1.26.

Chassis

Frame base of longitudinal and transverse members, welded to the self-supporting body. Independent suspension with torsion-bar springing and double-acting telescopic shock absorbers. Torsion-bar stabilizer on the front axle, additional rubber hollow springs on the rear axles. Ross steering with hydraulic steering damper, turning circle approx. 12 meters. Hydraulic four-wheel foot brakes with 1028 square cm effective braking area, mechanical hand brake on the rear wheels. Tubeless 700-14 tires. Fuel tank capacity 40 liters, reserve indicator via fuel gauge. Wheelbase 2400 mm, front and rear track 1375/1360 mm.

Body

Front-cab body in self-supporting all-steel construction. All windows of safety glass. Vent windows for draft-free ventilation in the cab doors; hinged rear side windows in the VW Kombi and Bus. Door locks on both sides. Cab interior (passenger and luggage space in the Bus) fully covered, insulated against drafts and heat loss. Adjustable single driver's seat with adjustable back, two-seat passenger bench. Two easily removable, upholstered three-seat benches in the VW Kombi for extra charge, standard in the Bus—also as two/three-seat sets. Regulatable ventilation system built into the roof for cab and passenger or freight compartment. Fresh-air heating with two closable heater openings in the foot space and two defroster ducts by the windshield, additional outlets according to body type, stepless fine regulation. Windshield washing system. Two padded sun visors. Handholds for passengers. Ashtrays, coat hooks. (Additional handholds, ashtrays, holding straps and coat hooks depending on model in the buses). Instruments including speedometer, odometer, fuel gauge (clock in VW Minibus "Special Model") as well as indicator lights for electricity, oil pressure, directional lights, high beams. Full-width shelf under the dashboard.

Weights and Measures

Greatest length x width x height a) for Box Van, Kombi and Minibus 4280 x 1750 x 1925 mm; b) for high-roof Box Van 4280 x 1750 x 2285 mm; c) for Pickup Truck and Double Cab 4290 x 1750 x 1910 (without top); d) for high roof Pickup Truck 4290 x 2020 x 1910 mm; e) for VW Minibus "Special Model" 4300 x 1800 x 1925 mm. Greatest width x height of load area doors a) double wing doors 1170 x 1200 mm; b) sliding door (extra charge) 1065 x 1205 mm; c) rear hatch 1230 x 730 mm (High roof box van 900 x 730 mm). Cargo space for Box Van or Kombi 4.8 cubic meters' in Bus without seats in passenger space 4.8 cubic meters, with seats in place 0.8 cubic meters in luggage space; in high roof box van 6 cubic meters. Rear bed of pickup truck 4.2 square meters; for high roof pickup truck 5.0 square meters; for double cab (only rear bed) 2.75 square meters. For all pickups except double cab, additional cargo space (lower lockup area) under rear bed, space 0.65 cubic meters. Allowable load according to model and equipment 910 to 1000 kg.

Performance

Top and sustained speed depending on body type 95/105 kph at 3660/4040 rpm. Climbing ability in 1st through 4th gears 28.0/14.5/8.0/4.5%. Fuel consumption according to DIN 70030*) depending on body type 9.7/10.0 liters per 100 km.
*With half of maximum load at steady 3/4 of top speed, add 10% to measured consumption.

Gut
transportieren –
bequem
reisen.

Gut transportieren.

Above and left: Pictures from a 1966 brochure. In that year 176,275 of the Transporter alone were built. Construction had meanwhile been moved to Hannover.

Bequem reisen.
Mit dem VW-Siebensitzer oder VW-Neunsitzer Sondermodell.

Travel comfortably.
With the VW Seven-seater or Nine-seater Special Model.

Production of the Bulli with divided windshield continued until June of 1967. Until then the whole production had reached the number of 1,853,439 vehicles—a success equaling that of the VW Beetle. The Samba Bus was still made until the 1967/68 model change.

en | Leergewicht 1070 kg[1] | Zul. Ges. Gew. 2070 kg
Nutzlast 1000 kg

Kastenwagen | Leergewicht 1110 kg[1] | Zul. Ges. Gew. 2070 kg
Nutzlast 960 kg

agen ohne/mit Plane und Spriegel | Leergewicht 1085/1120 kg[1] | Zul. Ges. Gew. 2070 kg
Nutzlast 985/950 kg

Holzpritsche | Leergewicht 1160 kg[1] | Zul. Ges. Gew. 2070 kg
Nutzlast 910 kg

Pritschenwagen | Leergewicht 1130 kg[1] | Zul. Ges. Gew. 2070 kg
Nutzlast 940 kg

Kombi | Leergewicht 1140 kg[3] | Zul. Ges. Gew. 2070 kg
Nutzlast 930 kg

Pritschenwagen m. Doppelkabine ohne/mit Plane und Spriegel | Leergewicht 1130/1150 kg[1] | Zul. Ges. Gew. 2070 kg
Nutzlast 940/920 kg[5]

Siebensitzer/Neunsitzer | Leergewicht 1150 kg[3] | Zul. Ges. Gew. 2070 kg
Nutzlast 920 kg

Siebensitzer/Neunsitzer Sondermodell | Leergewicht 1150 kg[3] | Zul. Ges. Gew. 2070 kg
Nutzlast 920 kg

Variant | Leergewicht 1025 kg[5] | Zul. Ges. Gew. 1400 kg[5]
Nutzlast 375 kg

bei 5 unbesetzten Sitzplätzen [3] einschl. Fahrer und Sitzeinrichtung

[5] gegen Aufpreis 465 kg, dann Gesamtgew. 1490 kg.

Transport well, travel comfortably. Why?
Because the whole program is a model of planning.

Box Van Dry weight 1070 kg Allowable gross weight 2070 kg Load limit 1000 kg

Kombi Dry weight 1140 kg 3) Allowable gross weight 2070 kg Load limit 930 kg

High Roof Box Van Dry weight 1110 kg Allowable gross weight Load limit 960 kg 2070 kg

Double Cab Pickup Truck -/+ cover & bows Dry weight 1130/1150 kg Allowable gross weight 2070 kg Load limit 940/920 kg

Pickup Truck -/+ cover & bows Dry weight 1085/1120 kg Allowable gross weight 2070 Load limit 985/950 kg kg

Seven-/Nine-Seater Dry weight 1150 kg 3) Allowable gross weight 2070 kg Load limit 920 kg

Wide Body Wooden Pickup Dry weight 1160 kg Allowable gross weight 2070 kg Load limit 910 kg

Seven-/Nine-Seater Special Model Dry weight 1150 kg 3) Allowable gross weight 2070 kg Load limit 920 kg

Wide Body Pickup Truck Dry weight 1130 kg Allowable gross weight 2070 kg Load limit 940 kg

Variant Dry weight 1025 kg 3) Allowable gross weight 1400 kg 3) Load limit 375 kg

3) 465 kg for extra charge, then gross weight 1490 kg.

The Bulli as Seen in the Press

The appearance of the VW Transporter, soon followed by a bus version, was observed in the press with almost exclusively positive comments. "The unbeatable characteristics of the Volkswagen car," it was said in the journal *Transport and Technology*, "along with the technical necessities of moving freight, resulted, with completely new exterior form, in a commercial vehicle that is desired on the German and foreign markets." The Transporter was designed and built with the most modern developments in mind, as a modern commercial vehicle, featuring a "usefulness as versatile as possible for all kinds of transport jobs".

"The vehicle is designed as a completely self-supporting all-steel construction with a very low, unbroken rear bed . . . The vehicle weighs 900 kilograms dry and carries 825 kilograms, thus offering a first-class weight/load ratio for a commercial vehicle of almost one to one." Fully loaded, the VW Transporter attained a top speed of 75 kph, could master upgrades of up to 22% (in first gear), and used 9.5 liters of fuel per 100 kilometers, on an average. "The outstanding handling characteristics of the passenger car, such as suspension, roadholding and acceleration, have been transposed to the eight-passenger bus version. This allows suspension settings that are simply not attainable with varying weights on the axles, and also results in fully even use of the tires' carrying ability and the brakes." The VW Bus was described as remarkably economical and "most highly useful", right to meet all kinds of customers' wishes. The interest of foreign customers was also stressed.

Many test reports were published in the motoring press in Germany and elsewhere. The Americans seemed particularly surprised: "An unbelievable vehicle, whose technical conception has been withheld from us far too long" *(Motor Trader)*.

"The second automobile revolution since the Beetle", or "Detroit will have to think of something", it was said *(Cars Illustrated, Motor Trend)*.

Even when the VW Bus had already been on the market a few years and one had long since become accustomed to seeing it on the street, new songs of praise were heard. The well-known motoring journalist Werner Buck wrote with true enthusiasm in *Auto, Motor und Sport* in 1957: "It is really fun to drive the eight-seater. The steering wheel is well-located, the central shift lever is easy to use . . . the steering proves to be completely shock-free, the clutch pedal works with the least application of power (about 15 kg), and the foot brake is praiseworthy too." Buck also spoke positively about the good roadholding and the high degree of driver safety.

Almost half of all small transporters at that time were VW Bullis, of which nearly 260,000 had left the assembly lines by the spring of 1957. "The newly opened factory in Hannover-Stöcken produces 300 of which 12,000 were special versions, seven- or eight-seaters. This last figure becomes impressive only when one knows that it represents 91% of the total German production of vehicles of this type."

Above: Wing doors were replaced by sliding doors in 1964 (optional a year earlier). Right: View of the motor compartment of a 1951 model. Below: Such loads could also be mastered.

Above: The Bulli as a mobile post office was often seen at fairs and exhibitions. Below: VW Pickup with turntable aerial ladder.

Right: Bulli in industrial use in 1964. Windshield washers were also standard from then on.

The success of the VW Bulli, as Buck wrote, could well be attributed to its being a combination of truck and car, "a means of transport for freight, of course, but also a representative of the trends that are customary in passenger car construction, often with the retention of complete passenger-car aggregates, least of all the motor. From this result good performance and pleasant driving characteristics." The Bulli's good reputation was also based on that of the VW motor, which is regarded as a model of reliability and longevity. "Also retained are the transmission and, in principle, the rear axle. But

Above: Whether in snow, in desert sand or on completely "normal" poor roads in South America, Sicily or Malaysia— with the Bus one was and is never stuck; it always came through.

Countless long trips and expeditions on all the continents were undertaken with the VW Bus. Many owners of such a vehicle rebuilt their vehicles themselves.

compared to the VW car, each rear axle ends in a spur gear that increases the ratio of the entire axle by 40%, thus suiting the power train to the particular nature of the Transporter. The principle of this double proportion is known from the earlier VW cross-country and amphibian vehicles. As in the passenger cars, the front wheels are carried on crank arms and sprung over torsion springs, naturally with all elements strengthened appropriately. Thus the spring unit has nine springs. Since 1955 the rear axle, still described as a 'shortened pendulum axle' as before, has had telescopic shock absorbers; here too we find transverse torsion bars. On the other hand, the body is built to be self-supporting without a frame with a central tunnel. This light construction consists of pressed sheet metal parts welded together; sidewalls and roof are strengthened by box sections, floor and inner walls by reinforcements. The axles are linked by a rigid frame of longitudinal and transverse links, resulting in a stiff unit that allowed the central tunnel to be dispensed with."

Buck used the "Samba Bus" as the object of the test and described the wealth of equipment to his readers. Nothing was spared to give comfort to the passengers; there was interior lighting, three armrests, an ashtray for each row of seats, a wide folding table on the back of the front seat, imitation leather covering the whole interior, handholds on the seat backs. Only the instrument panel was reduced to the most necessary. And this bus still had directional signal levers. "The directional signal is on the left side of the steering column, but it does not turn itself off.

There is no corner of the world which the VW Bus has not already reached...

Directional lights would be more modern, and certainly more practical in view of the vehicle's length. On the right of the steering column is the ignition-starter lock. Under the instrument panel, which has a clock at the far right and a place to install an optional radio in the center, there is a full-width shelf, somewhat flat but deep, well suited to atlases, maps, briefcases etc. Two other compartments, small ones to be sure, are located in the two front doors. The ashtray for driver and front-seat passenger is right near the windshield at the top of the dashboard. The ivory-colored steering wheel has two spokes, is very appropriate in terms of diameter and thickness, and has a horn button with the Wolfsburg arms in the middle. At the foot of the driver's seat, in the middle of the vehicle, is the heater control handle, left of it the starter control, at right the knob to activate the fuel valve.

"The Wolfsburg folks have done very much for ventilation, which can be regarded as excellent. The original VW heating system was retained, for warm air can be regulated without steps by pulling a handle, and a flap in the front ducting lets the warm air either come into the foot space or carries it on to the defroster outlets."

The front seat comfort drew criticism: "No wonder—when one sits right over the front axle that makes itself known with hard, short jolts, even when the vehicle is completely unloaded. Even stronger upholstery would be no mistake." And the imitation leather seat covers made one sweat. In addition, the seat back is not adjustable, nor is the seat itself . . . By driving thriftily, Werner Buck covered 100 kilometers on 8 liters of fuel; the average, including much city traffic, was 10.8 liters. This was respectably little in view of what a normal passenger car consumes.

"Fully loaded, the VW 8-seater has a power-to-weight ratio of a little over 60 kg/HP. On that basis, the acceleration figures are very favorable: standing start to 60 kph in 21 seconds (1st through 3rd gears), to 80 kph in 48 seconds. The top speed, also with 1850 kg gross weight, and with a partly opened roof panel, was 88 kph. On the windshield there is a warning label advising that the vehicle should be driven at no more than 80 kph, and red marks on the speedometer correspond to this advice. In practice this is not taken so seriously, as every superhighway driver knows from firsthand observation. The factory wants to protect itself thereby against overrevving the engine, and whoever values a long vehicle life and economical running will not choose any higher top speed on a flat superhighway. It is thus recommended to check the speedometer, since 80 kph there is by no means a genuine 80 kph (75 kph on the test vehicle). As end points for the gears, 15, 30, 50 and 80 kph are indicated, and they can be exceeded by quite a lot, which does not add anything worth mentioning to the acceleration, since the high point of the power curve has definitely been passed by then. I found the elasticity very considerable, for fourth gear can, if necessary, be used down to 15 kph, and third gear reaches from 10 to 70 kph if necessary. The locking synchronized transmission proved again to be a pure joy. A climbing ability of 4% in top gear is not much, but third gear practically doubles it, and it is always usable from 60-65 kph down. The full vehicle can also be driven spiritedly and quickly, with fast shifting."

Many a fire vehicle was in service for a long time. This VW from Hachenburg-Wörsbach dates from 1951.

with Full Room and Board". Since March 1 there was officially a VW Camper, a "one-room comfort dwelling", requiring a lot of mechanical and hand actions and even more camping experience to turn the VW Kombi into a usable home. "As easy as you have always imagined it to be, it definitely is not.

"Park your car on the parking lot in summer and come back an hour later. Then you will know that one cannot live in a body that stands at Lake Garda without a strong wind.

"That is why the VW Camper is double-walled. But unfortunately not doubly insulated. Which is why I would not care to comment on the plywood interior at this time. I have not yet slept in the vehicle, not that it is short, but that I wanted a test vehicle.

"When it gets too warm for one, one must simply leave the roof hatch and this or that window open. And when one returns to a parked, closed vehicle, one will have to open the doors carefully and then run away fast until the hot air has come out of it."

Fritz B. Busch listed VW Bus users or those whom he advised to make friends with the vehicle: "Business travelers, such as journalists (with darkroom and writing room in the back), sales people, exhibitors (who travel from fair to fair), expedition members, adventuring writers and business families, who always regretted that their Kombi stood around unused on Sundays."

All of them could now live in their vehicles over the weekend and on holidays. "There I see the greatest chance for this new-made vehicle—that one can use it in two ways, as a business car on working days and a caravan on vacation."

In the spring of 1961 the testers of *Auto, Motor und Sport* again took on the VW Bulli. This time it was the well-known motoring journalist Fritz B. Busch who described the vehicle, a camper. The title: "VW

Thousands of Camper Bullis were sold, and many box vans and Kombis were privately rebuilt and expanded. Secondhand government vehicles (mail vans) and veterans with several hundred thousand kilometers behind them thus enjoyed their second spring. In 1965 the magazine *Twen* held a contest for caravan owners who had built their own vehicles. Among the barely 150 entrants who presented their creations, there were—one scarcely could have expected anything else—more than 100 VW Bulli owners. Artur Westrup of the Volkswagen PR Service "Autopress" showed his professional colleagues many more rebuildings later: double-deckers (with a complete second Bulli body as a second floor!), eight-seaters cut down to convertibles, and particularly long or even short versions—thought up for whatever particular use. What the Netherlands VW importer Ben Pon had passed on to the Wolfsburg folks as his "intoxicated idea" developed into one of the most useful and universal vehicles in the world.

VW Bullis in all types of superlatives: the shortest bus served an Austrian community for years as a snowplow, the longest bus (with special roof construction) was seen some time back on a European tour in Hamburg, and a doubledeck bus—perhaps not the only one—could also be seen, in Amsterdam for example. The handling characteristics of all these vehicles may well have been very varying . . .

Development 1950-1967

1950 Box van production begins in February, Kombi in March, Bus in May. Motor: 1131 cc, HP at 3300 rpm.

1951 Samba Bus production begins in June, Ambulance in December.
1952 Open truck production begins in August. Second to fourth gears are synchronized.

1953 New 6-volt 84 Ah battery.

1954 January: 30 HP 1192 cc motor, 1:6.1 compression ratio, grooved pistons. August: compression ratio raised to 1:6.6, flat pistons.

1955 All models fitted with full-width dashboard. The spare wheel is now behind the driver's seat. The fuel tank is moved to over the rear axle on the right side (in closed models). Hydraulic steering dampers. Smaller 15-inch wheels.

1956 Production moved from Wolfsburg to Hannover.

1959 Introduction of the pickup truck with double cab. The crankshaft is strengthened, the crankcase enlarged. The exhaust system is changed, the gears fully synchronized. Stronger bumpers.

1960 The directional levers are replaced by lights. Motor now gives 34 HP at 3600 rpm. Automatic starting.

1962 Enlarged foot space via flat headlights. The passenger seat bench now folds forward, the driver's seat and back are adjustable. Optional 42 HP motor. Larger wheel wells.

1963 Box van with sliding side door optional. The one-ton version with 1.5 liter motor is added to the program. Roof hatch and rear hatch are enlarged considerably, front directional lights enlarged.

1964 Seat covers are now patterned. Larger and stronger windshield wipers. Improved heating and ventilation. Engine speed governor.

1965 Larger valves now give 44 HP for the 1.5 liter motor. Two-speed windshield wipers. Foot dimmer switch eliminated. Motor hood latch is a push button without lock. Open trucks get bigger rear windows. End of 34 HP Transporter production.

1966 Third gear ratio is lowered. Twelve-volt electric system becomes standard. Large elastic shift knob. One-key system. Optional inclusion of seat belt attachments.

1967 Production of the VW Bulli with split windshield ends.

Wendax
FRONT
★ TYP WL 1200
1¼ TONNER

AUS DEM WENDAX-PRODUKTIONS-PROGRAMM 1949

Right: as early as October of 1949, even before the official introduction of the VW Transporter, which was presented four weeks later, series production of the DKW Quick-Loader began at Auto Union. The vehicle was based on the well-known DKW Meisterklasse and had its two-cylinder two-stroke motor as well as its front drive.

Wendax

FRONT
Type WL 1200
1 1/4 Ton
From the Wendax production line of 1949
Wendax
The Dimensions
The Price DM 7760.00

The Wendax firm in Hamburg originally specialized in building light trucks. In 1949-50 they also built this transporter—with used Volkswagen motors! After building about a dozen examples, they gave up the experiment.

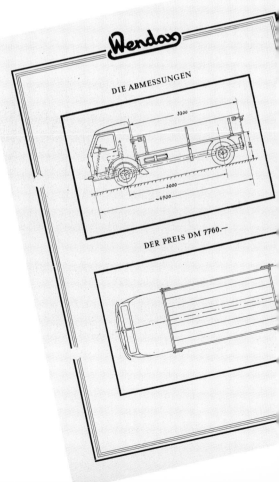

Wendax

DIE ABMESSUNGEN

DER PREIS DM 7760.—

DKW -Schnell-Laster 3/4 to

DKW Quick-Loader
3/4 ton

DKW Quick-Loader 3'6
DKW Safety

DKW
SCHNELLASTER
3=6

DKW = Sicherheit

AUTO UNION G·M·B·H

DKW
Fast Delivery Truck
3/4 Ton

The DKW produced at first only 20 HP—in this respect the VW Bulli was superior to it from the start. This is a 1950 brochure.

AUTO UNION

From 1949 to 1962, 58,792 trucks of this type were built in Ingolstadt; only a few have survived.

Available as box van and delivery truck.

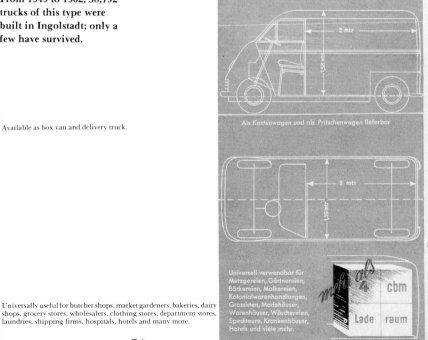

Als Kastenwagen und als Pritschenwagen lieferbar

Universell verwendbar für Metzgereien, Gärtnereien, Bäckereien, Molkereien, Kolonialwarenhandlungen, Grosslisten, Modehäuser, Warenhäuser, Wäschereien, Speditiere, Krankenhäuser, Hotels und viele mehr.

Universally useful for butcher shops, market gardeners, bakeries, dairy shops, grocery stores, wholesalers, clothing stores, department stores, laundries, shipping firms, hospitals, hotels and many more.

TECHNISCHE DATEN UND ABMESSUNGEN

MOTOR

Bohrung und Hub	mm	76 x 76
(2-Takt, 2-Zylinder)		
Hubraum	ccm	690
Leistung	PS	20
Verdichtung		1 : 5,9
Schmierung: Beimischung von Öl zu Benzin		1 : 25

FAHRZEUG

Radstand	mm	2500
Spurweite	mm	1190
Eigengewicht	etwa kg	850
Nutzlast	kg	750
Zulässiges Gesamtgewicht	kg	1700
Bereifung		5,50 x 16

Brennstoffverbrauch bei gleichbleibender Geschwindigkeit von

50 km/h	l/100 km	8-9
Höchstgeschwindigkeit	km/h	65-70
Bremsen - hydraulisch		4-Rad
Wendekreis-⌀	etwa m	11

Laderaum:

Länge	mm	2000
Breite des Kastens	mm	1390
Breite der Pritsche	mm	1500
Höhe (Kastenwagen u. Pritsche mit Plane)	mm	1580
Kastenwagen-Nutzraum	m³	4,2
Höhe der Ladefläche über der Fahrbahn (beladen)	mm	300
Gesamtlänge des Fahrzeuges	mm	3870
Breite des Fahrzeuges	mm	1550/1600
Höhe des Fahrzeuges	mm	1895

AUTO UNION G. m. b. H.
Filiale München
München 2, Albrechtstraße 14 - 20
Telefon: 36 27 68 u. 60 8 29

Änderungen vorbehalten - Abbildungen sind unverbindlich   D 102 (44950)

TECHNICAL DATA AND DIMENSIONS

MOTOR
Bore & stroke (mm) 76 x 76
(2-stroke, 2-cylinder)
Displacement (CC) 690
Power (HP) 20
Compression ratio 1: 5.9
Lubrication: ratio of oil to gasoline 1: 25

VEHICLE
Wheelbase (mm) 2500
Track (mm) 1190
Net weight (kg) 850 (approx.)
Load limit (KG) 750
Gross weight (KG) 1700
Tires 5.50 x 16

Fuel consumption at steady 50 kph speed 8-9 lit
Top speed (kph) 65-70
Brakes—hydraulic 4-wheel
Turning circle (m) 11 (approx.)
Load space:
Length (mm) 2000
Box width (mm) 1390
Pickup width (mm) 1500
Height (box van & pickup w/ cover, mm) 1580
Box van usable space
(square meters) 4.2
Bed height above road surface (loaded, mm) 300
Overall length (mm) 3870
Width (mm) 1550/1600
Height (mm) 1895

TAUNUS TRANSIT

¾-Tonner · 1-Tonner · 1¼-Tonner

TAUNUS TRANSIT
3/4 Ton—1 Ton—1 1/4 Ton

K-SM278

Only in March of 1953 did the Ford factory in Cologne follow with the Transit. To be sure, they offered motors of 38 to 60 HP. This competitor of the VW Transporter—soon also available in a variety of models—numbered a quarter million by 1965.

75

Tempo *Matador*

Der schnelle, wirtschaftliche 1-Tonner besitzt ein robustes Fahrgestell und ist durch seine Kurzbauform außerordentlich wendig. In seiner Luxus-Ausführung zeichnet er sich bei größter Geräumigkeit durch besondere Eleganz aus. Für die verschiedenen Transportzwecke ist der Matador mit Hochladerpritsche, Tiefladerpritsche, Kasten und Spezialaufbauten lieferbar.

The fast, economical 1-ton truck has a robust chassis and is extraordinarily nimble because of its short construction. In its luxury version is stands out through its particular elegance along with its great roominess. For various transport purposes the Matador is available with ligh-load bed, deep-load bed, box and special bodies.

At the same time as the VW Bulli, the Tempo Matador came on the market. It likewise had a VW motor as its powerplant, installed behind the front axle. Until the summer of 1952, 14,000 of the not very handsome Tempo trucks were built. As the little brother of the Matador there was the Wiking, a 3/4-ton truck with a 450 cc Heinkel motor, producing a modest 17 HP . . .

Tempo Wiking
The economical, high-performance 3/4-ton truck

Der preiswerte leistungsstarke ¾ Tonner

The further development of the Wiking Rapid, powered from 1957 on by a 32 HP Austin motor, could be called a lucky gamble. This model was produced untio 1963.

WIKING RAPID 32 PS | 4 ZYLINDER 4 TAKTER

WIKING RAPID 32 HP 4-CYLINDER 4-STROKE

A variation on the VW Samba Bus theme! To be sure, the Wiking had a considerable luggage space where the Bulli's motor sat.

Gutbrod *Atlas* 800   GANZ WIE SEIN NAME SAGT: EIN GUTER, STARKER HELFER!

Gutbrod Atlas 800 Just as its name says, a good, strong helper!

Some 11,000 of the Gutbrod Atlas were made from 1950 to 1954. The two-stroke, two-cylinder motor existed in 800 and 1000 cc versions, producing 16 and 18 HP respectively. Westfalia of Wiedenbrück produced various bodies for the Atlas. Professor Max Reisch undertook an expedition to Arabia with such a truck in 1950.

The successor to the Goliath GV
800 was available with 29 or 40 HP
motor (left) and could be regarded
as a really useful transport truck.
The "Express" was built until
1961 as a Kombi, box van, pickup
and bus.

Goliath GV 800, also a rival of the Bulli.
This vehicle too was modestly powered—
with a 450 or 600 cc motor, 16 or 21 HP.
About 4000 were built between 1951 and
1953.

By its construction and performance, but not least because of its
appropriate body types, the **Goliath 4-wheel Truck GV 800 A** best
suited to the most various transport tasks. It is **bigger, faster, stronger**
than you would expect from a truck of this class. Its considerable
reserves make themselves known pleasantly at every opportunity, be it
in city traffic, on long stretches or with extraordinary loads. All in all,
a truck whose development was influenced decisively by long years of
experience in building commercial vehicles and the awareness of
successful driving with the 19-time Goliath world record truck. A
truck you will take pleasure in!

The undemanding 21-HP two-stroke motor with 600 cc displacement
gives the truck outstanding acceleration and a top speed of
approximately 75 kph. The weight distribution—and thereby also the
handling and suspension characteristics—of the four-wheel truck are
outstanding for the empty or loaded truck.

The four-wheel box van has a cargo space of 2800 x 1390 x 1200 mm
and can be loaded easily through the double rear doors and the side
door.

The pickup truck with a rear bed of 2900 x 1620 mm is especially
desirable because of its manifold usefulness.

79

LLOYD LT 500

The Borgward firm's Lloyd marque was also represented by a transporter (box van, pickup, bus). But the half-ton truck was hopelessly underpowered with its 13 HP motor, and like the Lloyd firm's first passenger cars, the LT 500 also had a plywood body covered with imitation leather. In 1955 there appeared a 19 HP version, and sheet-metal bodies gradually followed. Total production of the LT 500: 24,668 units.

LLOYD LT Box Delivery Van

The most economical four-wheel half-ton van with the lowest price is a thoroughly handy and reliable fast transporter whose thriftiness is downright proverbial. It also offers the smallest businesses the possibility of utilizing the advantages of their own delivery van.

The 4.5 cubic meter inside space, with its extremely low cargo bed lying only 32 cm above the ground and, above all, being completely flat, allows full utilization to the farthest corner. Even heavy goods that are loaded all the way forward can be removed wothout trouble. The wide (900 mm), wide-opening rear door assures quick and easy loading and unloading.

As in a passenger car, the easy entry is behind the front wheels, considerably easing the driver's quick entry and exit in heavy street traffic. The arrangement of the entire powerplant, including fuel tank, in front of the driver's seat increases his safety and keeps the vehicle absolutely free from gas and oil fumes.

The smooth surface of the steel body on all sides allows any kind of advertising lettering. The driver's seat is equipped with all the comforts of the LLOYD passenger cars, including the standard built-in heater included in the price. At your request the LLOYD LT box delivery van will be delivered with or without a large-surface rear window.

Cargo space 3 cubic meters

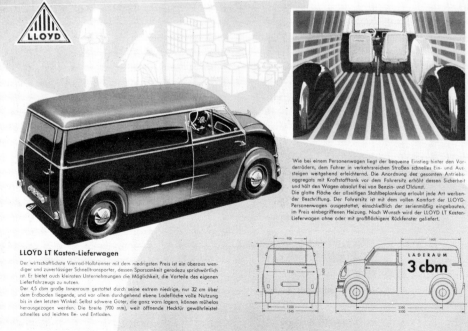

LLOYD LT Kasten-Lieferwagen

Der wirtschaftlichste Vierrad-Halbtonner mit dem niedrigsten Preis ist ein überaus wendiger und zuverlässiger Schnelltransporter, dessen Sparsamkeit geradezu sprichwörtlich ist. Er bietet auch kleinsten Unternehmungen die Möglichkeit, die Vorteile des eigenen Lieferfahrzeugs zu nutzen.

Der 4,5 cbm große Innenraum gestattet durch seine extrem niedrige, nur 32 cm über dem Erdboden liegende, und vor allem durchgehend ebene Ladefläche volle Nutzung bis in den letzten Winkel. Selbst schwere Güter, die ganz vorn lagern, können mühelos herausgezogen werden. Die breite (900 mm), weit öffnende Hecktür gewährleistet schnelles und leichtes Be- und Entladen.

Wie bei einem Personenwagen liegt der bequeme Einstieg hinter den Vorderrädern, dem Fahrer in verkehrsreichen Straßen schnelles Ein- und Aussteigen weitgehend erleichternd. Die Anordnung des gesamten Antriebsaggregats mit Kraftstofftank vor dem Fahrersitz erhöht dessen Sicherheit und hält den Wagen absolut frei von Benzin- und Öldunst. Die glatte Fläche der allseitigen Stahlbeplankung erlaubt jede Art werbender Beschriftung. Der Fohrersitz ist mit dem vollen Komfort der LLOYD-Personenwagen ausgestattet, einschließlich der serienmäßig eingebauten, im Preis einbegriffenen Heizung. Nach Wunsch wird der LLOYD LT Kasten-Lieferwagen ohne oder mit großflächigem Rückfenster geliefert.

Like scarcely any other vehicle, the box van offered itself as a logo carrier. The Volkswagen firm published a brochure in 1959 that showed numerous logo examples—to encourage many businesses.

WHO DRIVES A VW TRANSPORTER?

Some examples from the 1959 brochure—classic
examples of the commercial art of those times. Note
the advertisement for the E 605 poison—presumably
it was only seen for a short time . . .

VW Bus and Kombi Models

The VW Bulli is among the most commercial vehicle models: all the important German—and foreign—model manufacturers produced one or more versions. In view of the number of versions, only the most important manufacturers will be listed here—the Danish firm of Tekno alone produced over fifty body and logo variations. The most complete assortment was offered by the Berlin firm of Wiking, which introduced a VW Van as early as 1950 (Wiking #T9b), and their 1/87 scale models matched the contemporary real versions into the Eighties. The number given to one type usually remained unchanged; the most important basic models are listed with their numbers:

WIKING 1/87

29	Pickup truck
29 n	Pickup truck with cover
30	Delivery van
30 p	Mail van
29 d	Double cab truck
29 m	Double cab box van
31	Minibus
31 s	Special bus
31 p	Pickup truck (as of 1967)
32	Ambulance (van & Kombi)
60 k	Fire dept. van
60 u	Fire dept. ambulance
80 k	Oil heat service van
103	Police van

WIKING 1/40

The plastic display models for the Volkswagen firm were made in numerous versions and not available in stores.

SIKU 1/60

Bus, Kombi, pickup truck, ambulance of plastic, police van and radio van of metal.

MÄRKLIN 1/45

Bus and Kombi, sometimes with logo. Also an ambulance version.

MÄRKLIN 1/87

Pickup truck as railroad platform accessory.

GAMA 1/43

Bus, Kombi and pickup truck, fourteen versions in all.

DUX 1/38

Pickup truck and bus in metal or plastic, later in kit form.

TIPP & CO/TIPPCO ca. 1/20

Nice sheet metal models with flywheel motors, in five versions: Mail van, special bus with roof hatch, fire ladder truck, wrecker and Coca-Cola delivery truck.

Sheet metal models were made by many firms that no longer exist today: examples are the clockwork cars of GOSO or Philipp Niedermeier and CKO (Kellermann), or the flywheel cars made until the Seventies by Marchesini of Italy.

The best-known VW Bus miniatures from outside Germany came from Tekno (Ì3) and Lego (1/72, later Í7) of Denmark, and from Corgi Toys (five Ì3 versions) and Lesney Matchbox (1/75 delivery van and 1/64 camper) of England.

Less widespread were those by the English firm of Budgie Toys (1/76), DCMT Roadmaster Lone Star (1/60), Morestone (1/86) and Husky (1.75). Revell

made a nice 1/25 scale kit model of the Samba Bus in the USA—and from Brazil came two versions by Roly Toys in 1/60 scale.

In 1985 the German firm of Brekina made an excellent 1/87 model of the VW Bus (# 3100), costing only a fraction of what the historic models cost.

The Volkswagen Works not only gave assistance with painting and lettering the Bulli's side panels. They also published a thorough brochure that offered numerous examples of arrangement.

VW Transporter   purposeful interior layouts

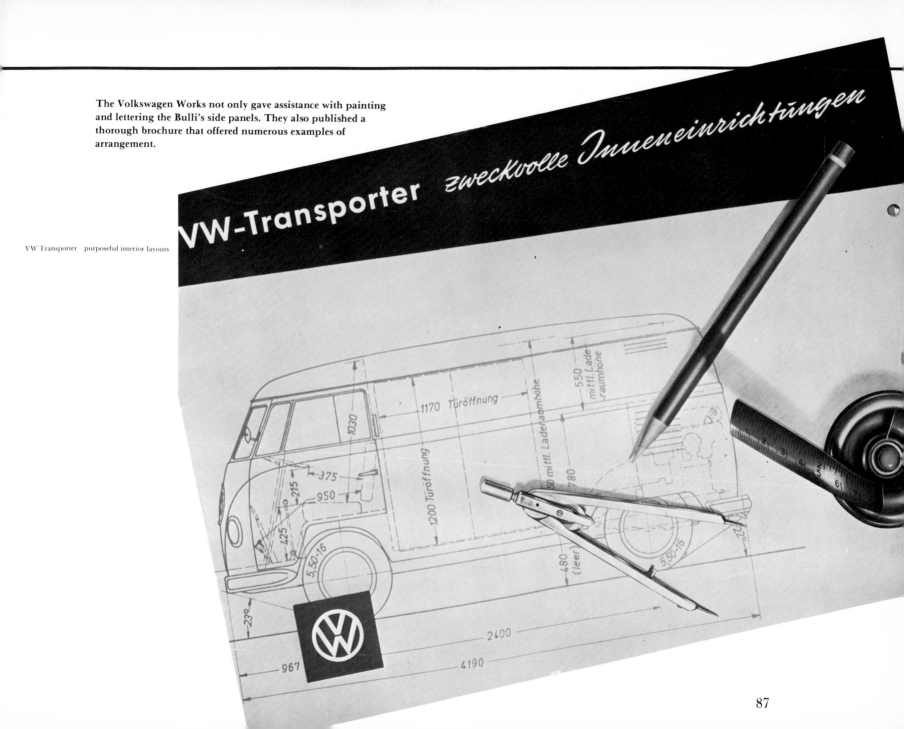

Examples for the construction of vending and delivery trucks. With
an added marquee the Bulli became the ideal vender's truck. Capable
carpenters prepared such trucks themselves too—much like VW
camping enthusiasts.

VW Box Van with interior shelves, sales counter and folding marquee
as mobile grocery store.

VW-Kastenwagen
mit Innenregalen, Verkaufstisch
und klappbarer Markise
als fahrender Verkaufsladen

VW-Kastenwagen
mit Innenverkleidung
sit-Hartfaserplatten (KW 504),
odenbelag, Rotationsentlüfter,
Regalen und Lattentrennwand
gsrichtung zum Brot-Transport

VW-Kastenwagen
mit Innenverkleidung aus Cefasit-Hartfaserplatten (KW 504),
2 Belüftern und Inneneinrichtung zum Transport von Brot,
Backwaren und Lebensmitteln aller Art

VW-Pritschenwagen
mit Kofferaufbau
und Entlüftungskiemen,
Zwischenregalen — seitlich
und hinten Flügeltüren —
zum Transport von Lebensr

VW Box Van with inner lining of Cefasite hardboard panels (KW 504), wood floor paneling, rotary exhaust fan, built-in shelves and longitudinal lath bulkhead for bread transport.

VW Box Van with interior lining of Cefasite hardboard panels (KW 504), 2 ventilators and interior set up to trans — port bread, baked goods and groceries of all kinds.

VW Pickup Truck with box body and vent flaps, inside shelves—side and rear wing doors—to transport groceries.

For grocery transport there were particular materials, insulating linings, ventilator flaps. Naturally the Bulli could also be turned into a refrigerator truck.

The VW Transporter as maid-of-all-work. Here VW proved that the Bulli could be utilized at airports: as a pilot car ("Follow me!", as a workshop vehicle, as an inspection car. Right: the Bulli as a radio car in the service of the Post Office. The VW Transporter also served as a radio transmitting station.

VW Box Van with racks for special oil and lubricant containers for the airport ground service crew.

VW Box Van with roof rack to hold a special ladder for the airport ground service crew.

VW Eight-seater as a special vehicle for the radio monitoring service, with extending mast.

VW-Achtsitzer
Spezialwagen für den Funkabhördienst
mit herausfahrbarer Peilanlage

VW-Achtsitzer mit Spezialanlagen und zusätzlichen Batterien als Funkwagen

VW Eight-seater with special equipment and additional batteries as a radio broadcast vehicle.

91

Patrol vehicle for the police—a well-known variation.
Incidentally, the first police radar vehicles were always VW Bullis.
A complete office setup made the Bulli into a mobile police
station. Right: technical data page from the end of 1959; the Bulli
still had directional levers instead of lights.

VW-Kombi als Verkehrs-Unfall-Bereitschaftswagen mit
Schreib-Klapptisch und Lampe sowie Funksprechanlage

VW Kombi as traffic as traffic accident
service vehicle with folding writing desk
and light as well as radio system.

VW-Achtsitzer
mit fest eingebauten Sitzbänken,
Klapptisch und Funksprechanlage
als Polizei-Bereitschaftswagen

VW Eight-seater with built-in seat benches,
folding table and radio speaker system as a
police patrol vehicle.

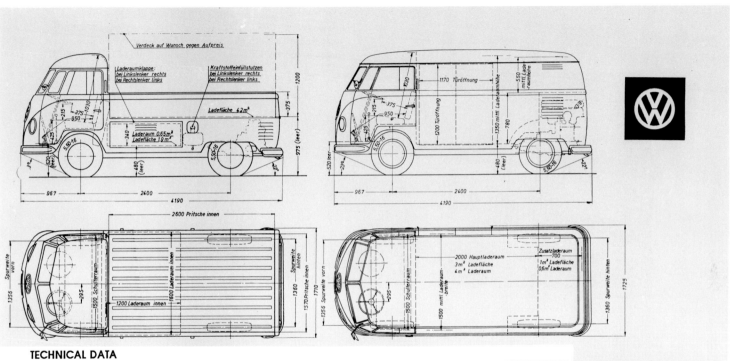

Verdeck auf Wunsch gegen Aufpreis.
Laderaumklappe: bei Linkslenker rechts bei Rechtslenker links
Kraftstoffeinfüllstutzen bei Linkslenker rechts bei Rechtslenker links
Ladefläche 4,2 m²
Laderaum 0,65 m²
Ladefläche 1,9 m²
2600 Pritsche innen
1600 Laderaum innen
1200 Laderaum innen
1500 Schulterraum
1570 Pritsche innen
1710
1360 Spurweite vorn
1356 Spurweite vorn
1360 Spurweite hinten

1170 Türöffnung
550 mittl. Laderaumhöhe
1350 mittl. Ladebanmhöhe
1200 Türöffnung
2000 Hauptladeraum 3 m² Ladefläche 4 m³ Laderaum
Zusatzladeraum 700 1 m² Ladefläche 0,6 m³ Laderaum
1500 mittl. Laderaumbreite
1725

TECHNICAL DATA

Weights	Pickup truck w/o cover	with cover	Box van	Kombi	8-seater 8-seater "Special Model"	Ambulance
	kg	kg	kg	kg	kg	kg
Net weight (tax weight)	935	935	900	925	1065	1185
Dry weight	1035 *	1070 *	1000 *	1025	1090	1210
Load limit	785	750	800	775	710	590
Gross wt.	1820	1820	1800	1800	1800	1800
Allowable axle weight (front)	1000	1000	925	925	925	925
Allowable axle weight (rear)	1000	1000	925	925	925	925
Number of seats	2-3	2-3	2-3	2-3 +	8	5-6

* Including driver, tools and spare wheel
+ With all seats: 8

Overall dimensions	Pickup truck w/o cover	with cover	Box van, Kombi, 8-seater	8-seater "Special Model"	Ambulance
	mm	mm	mm	mm	mm
Length	4190	4190	4190	4220	4190
Width	1710	1710	1725	1750	1725
Height (empty)	1890	2230	1900	1900	1900
Other dimensions					
Turning circle	ca.	ca.	ca.	ca.	ca.
	12 m	12 m	12 m	12 m	12 m
Ground clearance	250	250	250	250	250
Side hatch or door width	1125	1125	1170	1170	1170
Height (light)	400	400	1200	1200	1200

Interior dimensions	Pickup truck w/o cover	with cover	Box van, Kombi, 8-seater, 8-seater "Special Model"
Main cargo or passenger space (median measurement) in mm			
Length	2600	2600	2000
Width	1570	1570	1500
Height	375	1200	1350
Flr dist. above ground	975	975	480
Cargo surf.	4.2 m²	4.2 m²	3 m²
Cargo space	1.55 m³	1.55 m³	ca. 4 m³
Additional cargo or luggage space (median measurement) in mm			
Length	1200	1200	700
Width	1600	1600	1500
Height	340	340	550
Flr dist. above ground	480	480	1260
Cargo surf.	1.9 m²	1.9 m²	ca. 1 m²
Cargo space	.65 m³	.65 m³	ca. 0.6 m³

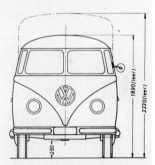

Addresses for the VW Bus Fan

Klub der Kafer-Freunde
W. Wölbling, Johann-Clanze-Strasse 49/V,
D 8000 Munich 70, West Germany

Volkswagen-Club Hamburg
Jürgen Ulmer, Klotzenmoor 55,
D 2000 Hamburg 61, West Germany

VW Uralt-Käfer-Club Schweiz
J. Eberhard, Tannenstrasse 14,
CH 9032 Engelberg, Switzerland

VW Brezelfenster-Club
Schulze-Delitzsch-Strasse 22,
D 3000 Hannover 61, West Germany

VW Veteranenklub Damnark
Mosevagn 105,
DK 5330 Munkebo, Denmark

VW Club Nederland
Bussumsestraat 96,
NL 2575 JM's Gravenhage, The Netherlands

Volkswagenhistoriska Klubben
Rohallsvagen 34,
S 18363 Taby, Sweden

Vintage Volkswagen Club of America
817 5th Street
Cresson PA 16630, USA

Society of Transporter Owners (SOTO)
P.O. Box 17234
Irvine CA 16630, USA

VW Split Window Club Great Britain
Bob Shaill, 194 Old Church Road,
St. Leonard-on-Sea, East Sussex TN38 9HD, England

A retired VW Bus as "tower" at a sporting airfield. As garden-house, tool shed or hot-dog stand, so many old Bullis have still performed outstanding service after being retired!

Das grosse Buch der Volkswagen-Typen: alle Fahrzeuge von 1934 bis heute by Lothar Boschen. This is the complete history of the Volkswagen up to 1983. All prototypes and production models are treated, with their variants. 588 pages, ca. 600 photos.

Die VW-Story by Jerry Sloniger. Everything about the development of the Beetle in historical and technical terms. The models are traced from the first prototype to the VW Golf and described in detail. 288 pages, 175 illustrations.

Der Bulli—Arbeit, Alltag und Kultur im Volkswagenwerk Hannover 1956-1984 by Herbert Flamme and Manfred Muster. The history of the VW Transporter from the Fifties till today. Likewise an accurate picture of working and living conditions of VW employees. Large format, 140 pages, many photos.

Alles über den VW-Bus: Wohnmobile—Reisemobile—Freizeitmobile—Modelle—Daten—Technik—Zubehör—Bausätze—Ausrüstung by E. Utz Orlopp and Martin Breuninger. The Bus is timelier than ever today and just as suitable as a camper as a city car. 14 x 20.5 cm, 250 pages, 150 illustrations.

Zwei Jahre im VW-Bus um die Welt by Stumpf Herb. A travel report with many practical experiences, technical tips and ideas for readers who long for distant places. 400 pages, 21 photos, 13 maps.

VW Transporter/Bus (50 PS) bis 5/79. How they're made, Etzold Vol. 17.

VW Transporter/Bus (68/70 HP) bis 5/79. How they're made, Etzold Vol. 18.

VW Transporter/Bus (1600/1700/1800/2000)/ Autobooks Repair Manual #733, in English.

VW Bulli im Gebirge. Four-color poster WK 1800, 50 x 70 cm.

Die deutschen Lastwagen der Wirtschaftswunderzeit—Band 1: Vom Dreiradlieferwagen zum Viereinhalbtonner by Bernd Regenberg. All model series of German truck production between about 1948 and 1959 are portrayed in words and pictures. In the first volume, all trucks with a capacity of up to 4.5 tons are presented, including all VW Transporters. 200 pages, 450 illustrations.

Lieferwagen, Transporter, Kleinbusse 1945-1980 by Werner Oswald. All vehicles are portrayed inclusively and completely with precise technical data and photos. 326 pages, 350 photos.

Double-cab at a building site. This body type was introduced in 1959 and was well received from the start.